Dam Engineering - Recent Advances in Design and Analysis

Edited by Zhongzhi Fu and Erich Bauer

Published in London, United Kingdom

IntechOpen

Supporting open minds since 2005

Dam Engineering - Recent Advances in Design and Analysis
http://dx.doi.org/10.5772/intechopen.78818
Edited by Zhongzhi Fu and Erich Bauer

Contributors
Liujiang Wang, Akira Murakami, Vikas Sharma, Kazunori Fujisawa, Qiming Zhong, Zhongzhi Fu, Shengshui Chen, Enyue Ji, Guoying Li, Yang Lu, Yijiang Zhang, Weiwei Xu, Yumeng Zhu, Daniel Dias, Xiangfeng Guo, Erich Bauer, Guoying Li, Zhankuan Mi, Kuangmin Wei, Yibo Shan, Jiaxin Liu

First published in London, United Kingdom, 2021 by IntechOpen
IntechOpen is the global imprint of INTECHOPEN LIMITED, registered in England and Wales, registration number: 11086078, 5 Princes Gate Court, London, SW7 2QJ, United Kingdom
Printed in Croatia

British Library Cataloguing-in-Publication Data
A catalogue record for this book is available from the British Library

Additional hard and PDF copies can be obtained from orders@intechopen.com

Dam Engineering - Recent Advances in Design and Analysis
Edited by Zhongzhi Fu and Erich Bauer
p. cm.
Print ISBN 978-1-83962-157-4
Online ISBN 978-1-83962-158-1
eBook (PDF) ISBN 978-1-83962-159-8

For EU product safety concerns:
IN TECH d.o.o., Prolaz Marije Krucifikse Kozulić 3, 51000 Rijeka, Croatia, info@intechopen.com or visit our website at intechopen.com.

Meet the editors

Zhongzhi Fu is a professor and senior engineer in the Department of Geotechnical Engineering, Nanjing Hydraulic Research Institute (NHRI), which is a non-profit research center governed by the Ministry of Water Resources, China. Dr. Fu specializes in hydraulic structures, especially for embankment and rockfill dams. He obtained his doctoral degree from Hohai University, Nanjing, China, in 2011 and became a consultant engineer for NHRI in June 2011. He has been working on testing and modeling the static and dynamic behavior of earth and rockfill materials and has published more than seventy technical papers in national and international journals and conference proceedings. Dr. Fu has participated in consulting works for more than thirty dams of different types, including the DaShiXia Concrete Faced Gravel Dam in Xinjiang (247 m in height) and the RuMei Earth Core Rockfill Dam in Tibet (315 m in height). He is now the secretary of the Soil and Rock Mechanics Committee in the Chinese Hydraulic Engineering Society and is a member of the Chinese National Committee on Large Dams.

Erich Bauer is a professor at the Institute of Applied Mechanics (IAM), Graz University of Technology, Austria. He completed his postgraduate studies at the University of Karlsruhe, Germany, under the supervision of Prof. G. Gudehus and Prof. D. Kolymbas, and holds a habilitation in Mechanics. Before joining IAM, he worked in a consulting company on research projects in Europe and the United States using the New Austrian Tunneling Method (NATM). Dr. Bauer is a regular reviewer of papers for various international journals, a Ph.D. thesis examiner, a keynote speaker at a large number of international conferences, and a member of different editorial boards. His scientific work focuses on the mechanical properties of granular materials and their constitutive modeling, using classical and higher-order continuum descriptions. His main areas of research are continuum mechanics, experimental investigations, and the constitutive modeling of engineering materials including the long-term behavior of geotechnical structures like rockfill dams. His scientific results have been published in 158 papers and books and presented in 136 oral presentations. Dr. Bauer is the project leader of several national and international scientific-technical initiatives with universities in Europe and Asia. He is also visiting professor at Hohai University, China, and the founder of the international conference series on “Long-term Behaviour of Dams” (LTBD).

Contents

Preface

Dams and reservoirs are significant parts of the infrastructure in many countries, because of their comprehensive functions of flood control, irrigation, power generation, environmental protection, and recreation. Deficiently constructed dams are sensitive to structural damages, which may lead to leakage and erosion, dam breaches, and environmental disasters. To ensure safe operation and avoid large costs to repair deficient dams, great efforts should be exercised in site investigation, designing, analysis, construction, and maintenance of the dam building.

In the past decades, many high dams have been successfully constructed worldwide. These active engineering practices could benefit greatly from advancements in construction technologies and enhanced designing theories. The engineering praxis, of course, also provides sufficient cases to summarize experiences and lessons learned about the long-term behavior of dams, and to evaluate the reasonability and reliability of designing theories, including those for deformation and stability analyses under different loading conditions, which can roughly be divided into static and dynamic loads, thermal, and hydro-chemical weathering.

This book presents recent research in both engineering practice and analytical methods in eight chapters. Each chapter contains valuable studies on specified areas of dam engineering. Chapter 1 presents properties of clay–gravel mixtures and the use of such gap-graded materials in four famous high rockfill dams in western China. Chapter 2 reports a case study on the heightening of an existing embankment dam from 63.4 m (height) to 98.1 m using a special impermeable protection system. Chapter 3 explains the theory of the space-time finite element method and its application in studying the responses of a concrete dam-reservoir system to earthquake loading. Chapter 4 presents a novel hypoplastic constitutive model for moisture-sensitive, coarse-grained rockfill materials based on the concept of degradation of the solid hardness. Chapter 5, within the framework of critical state theory, discusses a state-dependent elastoplastic constitutive model for rockfill materials, which is especially useful in predicting the complex deformation behavior of rockfill dams. Chapter 6 focuses on the efficient probabilistic analyses of embankment dams and introduces a practical framework in such a way that a first assessment stage using random variables (RV) for the soil variability is followed by a second assessment stage using random fields (RF). Chapter 7 is also about the probabilistic analyses of embankment dams, and the authors demonstrate the use of some well-known analytical methods in performing deterministic and probabilistic slope stability analysis for embankment dams. Chapter 8 provides a comprehensive review of recent research progress in the numerical modeling of the breach process of earth-rock dams.

We hope that this book provides an opportunity for readers working in relevant areas to utilize recent practical experiences and scientific findings.

Dr. Zhongzhi Fu
Professor,
Nanjing Hydraulic Research Institute,
Nanjing, P. R. China

Dr. Erich Bauer
Professor,
Graz University of Technology,
Graz, Austria

Chapter 1

Using Clay-Gravel Mixtures as the Impervious Core Materials in Rockfill Dams

Zhongzhi Fu, Shengshui Chen, Enyue Ji, Guoying Li and Yang Lu

Abstract

Constructing the impervious system of an Earth Core Rockfill Dam (ECRD) usually needs a large volume of clay that may exhaust a huge area of farmland. One way to reduce the volume of clay to be filled is to use natural clay-gravel mixtures (CGM) or to add an appropriate percent of gravel materials into the clay and use the artificial clay-gravel mixtures as the impervious core materials. Using clay-gravel mixtures can also effectively increase the modulus of the core and reduce the differential settlement between the core and its adjacent rockfill shoulders, and thus alleviate the risk of occurrence of potential cracks within the core wall. The impermeability behavior of the compacted clay-gravel mixtures, however, has to be careful investigated and verified. In this chapter, four cases in using clay-gravel mixtures in constructing high ECRDs in China are reviewed, with attention focused on the engineering properties of clay-gravel mixtures and the construction and field quality control aspects of ECRDs using clay-gravel mixtures.

Keywords: rockfill dam, clay-gravel mixture, earth core, impermeability

1. Introduction

An Earth Core Rockfill Dam (ECRD) is a type of dam that uses a central or inclined clay core wall as the impervious system, while rockfill materials are used to construct the shoulders of the dam [1]. Information of some constructed high ECRDs around the world has been summarized by Zhang [2]. Normally, a good performance of an ECRD necessitates that the clayey soil used in its core meeting the following requirements [1, 3]: (a) The modulus of the compacted clay core should be high enough so that the arch effect between the core and the shoulders is not so evident as to result in horizontal cracks and hydraulic fracturing in the core. (b) The permeability of the clay core should be low enough so that the leakage after impounding does not exceed the expected quantity. Both requirements can be satisfied by proper gradation design and adequate compaction of the clayey soils used in the core wall [4].

In recent years, several high ECRDs have been constructed in western China and a common feature of these dams is the use of clay-gravel mixtures (CGM), either natural or artificially blended, as the impervious core materials. Control the mass content of gravel plays a central role in controlling the strength, deformation and permeability behavior of the mixtures obtained [2, 5]. For instance, increasing the content of gravel results in an increase in the stiffness of CGM and is beneficial to reduce the differential settlement between the core and shoulders. On the other hand, the permeability also tends to increase as the content of gravel increases, which results in a potential risk of seepage failure and unacceptable leakage. A good design practice, therefore, needs a balance between the impermeability and deformation behavior. In this chapter, four cases of using CGM in constructing high ECRDs are reviewed, with particular attention focused on their engineering properties and the relevant construction concerns and field control tests.

2. Basic information of case studies

The four ECRDs considered are named PuBuGou (PBG) [5], ChangHeBa (CHB) [6–8], NuoZhaDu (NZD) [2, 9], and LiangHeKou (LHK) [10–12], respectively.

Dam		PBG	CHB	NZD	LHK
Height (m)		186	240	261.5	295
Reservoir volume (10^8 m^3)		53.37	10.75	237.03	107.67
Foundation	**Type**	**Sand and gravel**	**Sand and gravel**	**Granite**	**Sandstone and slate**
	Thickness (m)	78	79	-	-
Crest	Length (m)	540.5	497.9	630.1	650.0
	Width (m)	14	16	18	16
Dam slope (H:V)	Upstream	1:2.0 and 1:2.25	1:2.0	1:1.9	1:2.0
	Downstream	1:1.8	1:2.0	1:1.8	1:1.9
Core wall dimensions	Crest width (m)	4.0	6.0	10.0	6.0
	Bottom width (m)	96.0	125.75	114.6	150.0
	Ups. slope (V:H)	1:0.25	1:0.25	1:0.2	1:0.2
	Dws. slope (V:H)	1:0.25	1:0.25	1:0.2	1:0.2
Core wall materials	D_{max} (mm)	80	150	120	75
	P (d < 5 mm)	39–54%	52–56%	60–70%	55%
	P (d < 0.075 mm)	19–26%	26–29%	25–40%	28–44%
	P (d < 0.005 mm)	3.6–8.3%	8–9%	≥8%	12–17%
	Permeability (cm/s)	≤10^{-5}	≤10^{-5}	≤10^{-5}	≤10^{-5}
	ρ_d/ρ_{dmax}	≥98%	≥97%	≥95%	≥97%
Filter thickness	Upstream (m)	4 m ×2	8 m ×1	4 m ×2	4 m ×2
	Downstream (m)	6 m ×2	6 m ×2	6 m ×2	6 m ×2
Seismicity	PGA (PE_{100} = 2%)	0.225 g	0.359 g	0.380 g	0.288 g
	PGA (PE_{100} = 1%)	0.268 g	0.430 g	0.436 g	0.345 g

Note: D_{max} = *maximum particle diameter; d = diameter of soil particles; P = mass percentage of soil particles;* ρ_d = *dry density;* ρ_{dmax} = *maximum dry density; PGA = peak ground acceleration;* PE_{100} = *probability of exceedance within 100 years; 1g = 9.81 m/s*^{2}.

Table 1.
Basic information of the PBG, CHB, NZD and LHK ECRDs.

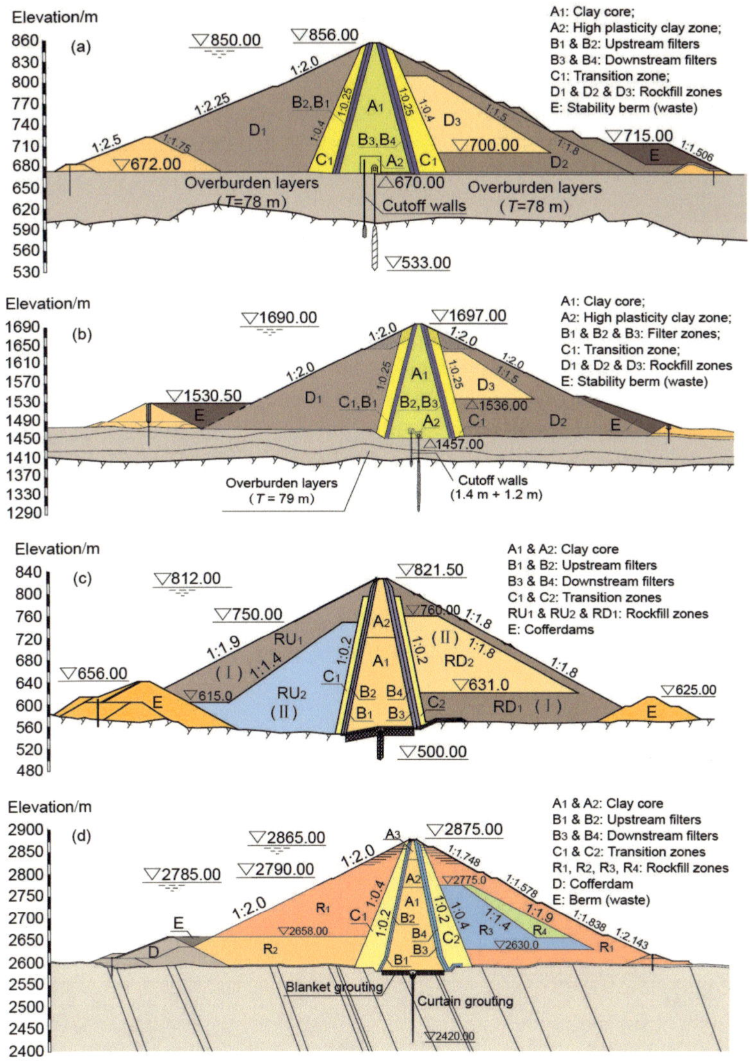

Figure 1.
Material zones of the four ECRDs. (a) The PBG dam; (b) the CHB dam; (c) the NZD dam and (d) the LHK dam.

All four dams use central clay core walls, protected by filter zones upstream and downstream. PBG and CHB are ECRDs constructed over thick overburden layers with maximum depth of about 80 m, while NZD and LHK are seated on rock foundations. The basic information of the four dams are summarized in **Table 1** and their zones of materials are shown in **Figure 1**, respectively.

3. Properties of clay-gravel mixtures

3.1 Compaction characteristics

Laboratory compaction test is the most fundamental experiment performed in geotechnical engineering [13, 14], which provides the basis for determining the percent compaction and water content needed to achieve the required engineering properties, and for controlling construction to assure that the required compaction and water content are achieved. For clay-gravel mixtures used in ECRDs, however, the existence of oversize fraction makes the two molds described by ASTM (with diameters of 101.6 mm and 152.4 mm, respectively) incapable of yielding reliable results. Therefore, special compaction molds with larger diameters were fabricated to establish the compaction curves. For instance, a modified compaction mold 500 mm in diameter was used for the PBG dam [5], and a mold with a diameter of 600 mm was fabricated for the NZD dam [2].

Figure 2 shows the influence of the mass percentage of soil particles larger than 5 mm (P_5) on the maximum dry density (ρ_{dmax}) and the optimum water content (ω_{opt}). Different compaction efforts were used for different cases. Note for the NZD dam P_5 was evaluated after compaction tests and for the rest three dams it was evaluated before compaction tests. Particle breakage may occur during compaction and the two approaches may give slightly different results. Nevertheless, common trends can be observed from **Figure 2**: an increase in P_5 from zero results in a steady increase in ρ_{dmax} until a threshold value is achieved, beyond which a further increase in P_5 leads to a rapid decrease in ρ_{dmax}. The value of this threshold is around 60–70% for the reviewed cases.

Corresponding to the different mass content of gravel particles, several states of clay-gravel structure can exist as illustrated in **Figure 3**, i.e., the state that gravel particles floating within the fine fraction, the state that gravel particles start to contact, and the state that fine fraction filling voids formed by gravel particles.

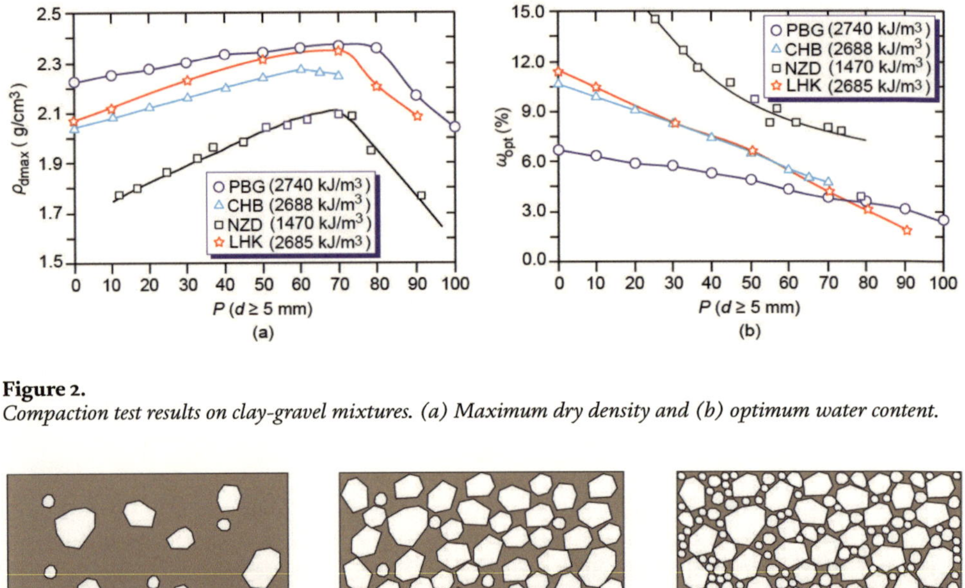

Figure 2.
Compaction test results on clay-gravel mixtures. (a) Maximum dry density and (b) optimum water content.

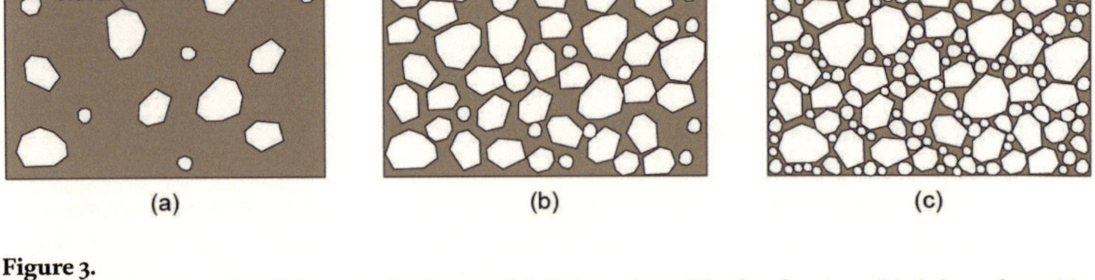

Figure 3.
Void filling characteristics of clay-gravel mixtures. (a) Skeleton formed by fine fraction, (b) skeleton formed by fine and gravel fractions, and (c) skeleton formed by gravel fraction.

In the first state the soil skeleton is formed by the fine fraction and the gravel particles seem to be floating within the soil matrix separately, i.e., only a very few or even no contacts are formed between gravel particles. In this case, the state of maximum dry density of the total material is achieved when the fine fraction itself is compacted to its maximum dry density. If we denote the maximum dry density of the fine fraction and the corresponding optimum water content by ρ^*_{dmax} and ω^*_{opt}, respectively, then the following theoretical relationship can be established for the maximum dry density of the total material (ρ_{dmax}):

$$\rho_{\mathrm{dmax}} = \frac{\rho^*_{\mathrm{dmax}} V_c + \rho_g V_g}{V_c + V_g} \tag{1}$$

in which ρ_g is the dry density of gravel particles. V_g and V_c denotes the total volumes of gravel particles and the fine fraction, respectively. The content of gravel, c_g, is defined as the mass of dry gravel divided by the total mass of the dry mixture, i.e.

$$c_g = \frac{\rho_g V_g}{\rho^*_{\mathrm{dmax}} V_c + \rho_g V_g} \tag{2}$$

which can be rewritten as follows:

$$V_g = \frac{c_g}{1 - c_g} \frac{\rho^*_{\mathrm{dmax}}}{\rho_g} V_c \tag{3}$$

Substituting Eq. (3) into Eq. (1) yields the following relationship:

$$\rho_{\mathrm{dmax}} = \frac{\rho_g}{\left(1 - c_g\right)\rho_g + c_g \rho^*_{\mathrm{dmax}}} \rho^*_{\mathrm{dmax}} \tag{4}$$

It can be verified that $\rho_{\mathrm{dmax}} = \rho^*_{\mathrm{dmax}}$ when $c_g = 0$ and ρ_{dmax} increases with c_g when $\rho_g > \rho^*_{\mathrm{dmax}}$. Various experiments have shown that the water content within the fine fraction is quite close to its optimum value (ω^*_{opt}) when the total material achieves its densest state [12, 15]. This conclusion can be used to establish the relationship between ω_{opt} and ω^*_{opt}, i.e.

$$\omega_{\mathrm{opt}} = \frac{\rho^*_{\mathrm{dmax}} V_c \omega^*_{\mathrm{opt}} + \rho_g V_g \omega_g}{\rho^*_{\mathrm{dmax}} V_c + \rho_g V_g} \tag{5}$$

in which ω_g denotes the water content of gravel particles when the CGM achieves a moisture equilibrium state and the fine fraction itself has a water content of ω^*_{opt}. Inserting Eq. (3) into Eq. (5) yields

$$\omega_{\mathrm{opt}} = \left(1 - c_g\right)\omega^*_{\mathrm{opt}} + c_g \omega_g \tag{6}$$

Obviously the optimum water content decreases almost linearly when the gravel content increases (**Figure 2**). In particular, the upper and lower bounds of ω_{opt} can be evaluated by setting ω_g to the water content that gravel particles are saturated but with dry surfaces and by setting it to zero, respectively.

As the mass content of gravel fraction (d > 5 mm) increases, the floating particles become increasingly close to each other and the theoretical threshold of this stage is that the gravel fraction achieves its loosest possible state, i.e., the particles just start to contact as shown in **Figure 3(b)**. In this state, the maximum void ratio of gravel fraction (e_{max}) can be expressed as follows:

$$e_{max} = V_c / V_g \tag{7}$$

Substituting Eq. (7) into Eq. (2) gives an empirical formula to estimate the upper bound of c_g that the floating pattern exist or the lower bound of c_g that the floating pattern starts to disappear, i.e.

$$c_g = \frac{\rho_g}{\rho^*_{dmax} e_{max} + \rho_g} \tag{8}$$

If the mass content of gravel fraction is further increased beyond the threshold given by Eq. (8), the soil skeleton will be formed by both the fine fraction and the gravel fraction. A higher c_g means a greater contribution made by the gravel fraction to the skeleton. In this case, the gravel fraction can be accommodated only by compacting the fine fraction to a looser state than the densest state, i.e., the fine fraction cannot be compacted to its densest state anymore. Correspondingly, Eq. (4) can be revised as follows:

$$\rho_{dmax} = \frac{\rho_g}{\left(1 - c_g\right)\rho_g + c_g \rho^*_d} \rho^*_d \tag{9}$$

in which $\rho^*{}_d$ denotes the dry density of compacted fine fraction. Two competitive trends are implicated in Eq. (9). For a given $\rho^*{}_d$, an increase in c_g still leads to an increase in ρ_{dmax}. However, the increase in c_g meanwhile results in a decrease of $\rho^*{}_d$. The combined effect is that the ρ_{dmax} of the total material continues to increase when c_g is increased, with, however, a different trend from the previous stage. A peak for ρ_{dmax} is achieved at a particular c_g, exceeding which the value of ρ_{dmax} decreases rapidly as shown in **Figure 5**. In this stage, the total material is still easiest to compact when the water content of its fine fraction is $\omega^*{}_{opt}$. It is easy to verify that Eq. (6) still holds.

The compaction characteristics can also be investigated from another extreme case that c_g = 1, i.e., the soil skeleton is completely formed by the gravel fraction. If the minimum void ratio of gravel fraction achievable after compaction is denoted by e_{min}, then the maximum dry density of the gravel fraction (total material in this case) reads:

$$\rho_{dmax} = \frac{\rho_g}{\left(1 + e_{min}\right)} \tag{10}$$

Adding a few amount of fine fraction will not change the soil skeleton, and the fine fraction is simply filling the inter-gravel voids. The dry density of the total material can then be expressed as follows:

$$\rho_{\mathrm{dmax}} = \frac{\rho_{\mathrm{d}}^{*} V_{\mathrm{c}} + \rho_{\mathrm{g}} V_{\mathrm{g}}}{V_{\mathrm{g}}\left(1 + e_{\min}\right)} \tag{11}$$

Because the compaction effort cannot be imparted to the fine fraction in this state, ρ^{*}_{d} can be interpreted as its dry density under uncompacted state. Inserting Eq. (3), with ρ^{*}_{dmax} replaced by ρ^{*}_{d}, into Eq. (11) yields

$$\rho_{\mathrm{dmax}} = \frac{\rho_{\mathrm{g}}}{c_{\mathrm{g}}\left(1 + e_{\min}\right)} \tag{12}$$

It is clear that an increase of clay content (decrease of c_{g}) results in an increase in ρ_{dmax}, and the threshold can be attained when all the voids are filled by fine fraction, i.e., $V_{\mathrm{c}} = e_{\min} V_{\mathrm{g}}$. In this case, Eq. (11) can be rewritten in the following form:

$$\rho_{\mathrm{dmax}} = \frac{\rho_{\mathrm{d}}^{*} e_{\min} + \rho_{\mathrm{g}}}{\left(1 + e_{\min}\right)} \tag{13}$$

Similar as Eq. (8), the threshold gravel content for this stage reads:

$$c_{\mathrm{g}} = \frac{\rho_{\mathrm{g}}}{\rho_{\mathrm{d}}^{*} e_{\min} + \rho_{\mathrm{g}}} \tag{14}$$

Since the fine fraction is not compacted in this state, the optimum water content of the total material (ω_{opt}) has nothing to do with that of the void-filling fine fraction ($\omega^{*}_{\mathrm{opt}}$).

If the content of fine fraction is further increased beyond the threshold given by Eq. (14), it will start to participate in forming the soil skeleton, and a lower c_{g} means a greater contribution made by the fine fraction. In this case, some compaction effort is imparted to the fine fraction and the gravel fraction cannot be densified to $e_{\min}$ anymore. Eq. (9) has already been established to describe the compaction behavior for this state. A decrease in c_{g} results in a less compacted structure for the gravel fraction. On the other hand, the decrease in c_{g} also leads to an increase in ρ^{*}_{d}. The influence of c_{g} on ρ_{dmax} depends on the two competitive effects. **Figure 4** depicts the compaction behavior of clay-gravel mixtures in three states.

3.2 Permeability

It is important to verify that the impermeability of the compacted CGM meets the requirement of the design code. The permeability coefficient of the clay core is usually controlled below 10^{-5} cm/s [1]. **Figure 5** shows the influence of gravel content on the permeability coefficient. The seepage experiments on the material for the LHK dam is performed by Lu with a low percent compaction so

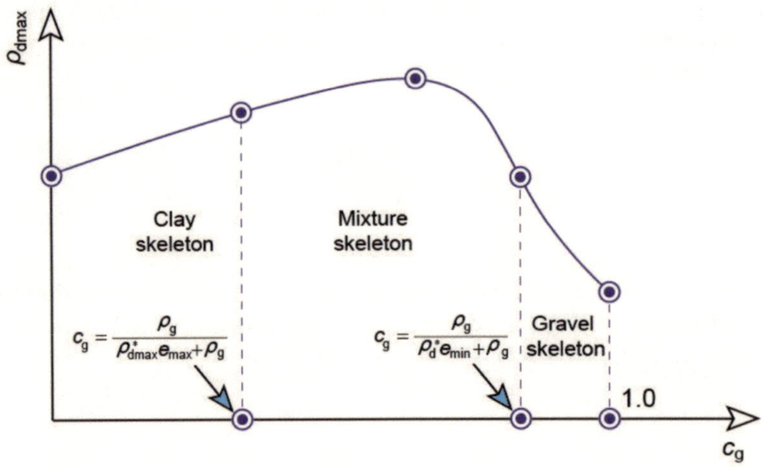

Figure 4.
Influence of c_g *on* ρ_{dmax} *in three skeleton states.*

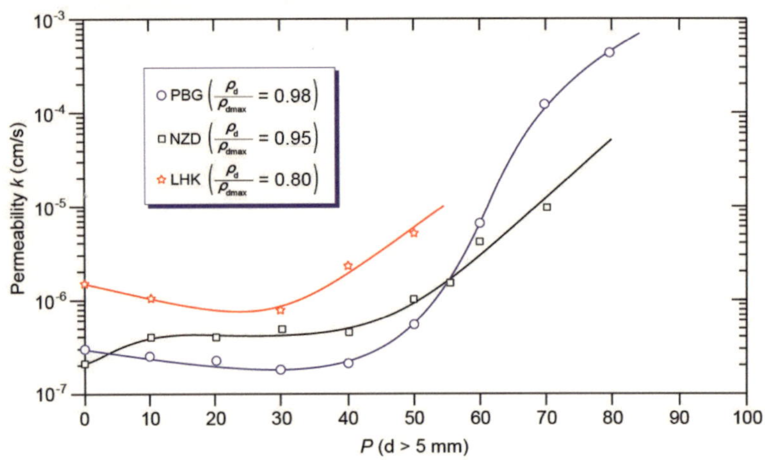

Figure 5.
Influence of gravel content on the permeability.

that particle breakage was expected not evident [12]. A common feature of the approximating curves in **Figure 5** is the almost constant or a slight decrease of the permeability when the gravel content (P_5) is increased from zero to about 30%. The lowest permeability can be achieved when P_5 is around 30%. Beyond this amount, the permeability coefficient increases rapidly with a further increase in P_5.

Abundant double-ring infiltrometer tests were performed during test filling of the CGM used in the PBG dam [5]. The rate of vertical infiltration (k_v) is plotted against the dry density (ρ_d) and gravel content (P_5) in **Figure 6**. The results are rather scattered, indicating the inhomogeneous nature of the filled materials. However, it is clear that the places where the rates of infiltration are relatively high ($k_v > 10^{-5}$ cm/s) are either not well compacted with a relatively low density or have relatively high gravel contents. Therefore, it is important to control the gravel content below a certain limit to ensure the impermeability of the compacted CGM. When the CGM is in a mixture skeleton state (**Figure 4**), embedding an increased amount of gravel particles leads to a decreased percent compaction of the fine fraction as pointed out previously. This tends to increase the permeability of the mixture. On the other hand, the embedded particles serve as seepage barriers as the permeability of gravel particles is considerably lower than the fine fraction. These two competitive effects control the dependence of the permeability of the total material upon the gravel content. Note that the rate of infiltration obtained in situ could not be used to determine the coefficient of permeability directly although both quantities have the same dimension [16].

Initial water content also has an influence on the permeability of compacted CGM. **Figure 7** shows typical results obtained from laboratory experiments with PBG CGM [5]. For a given grain size distribution and water content, the coefficient of permeability (k) decreases when the dry density of the total material is increased. However, the rate of decrease in permeability also decreases when the dry density is increased, indicating an increasingly difficulty in reducing the permeability. For the given compaction effort applied, the maximum dry density of the total material was achieved when ω = 4.625%. However, the permeability coefficient does not reach the minimum at this optimum water content: compacting the CGM slightly

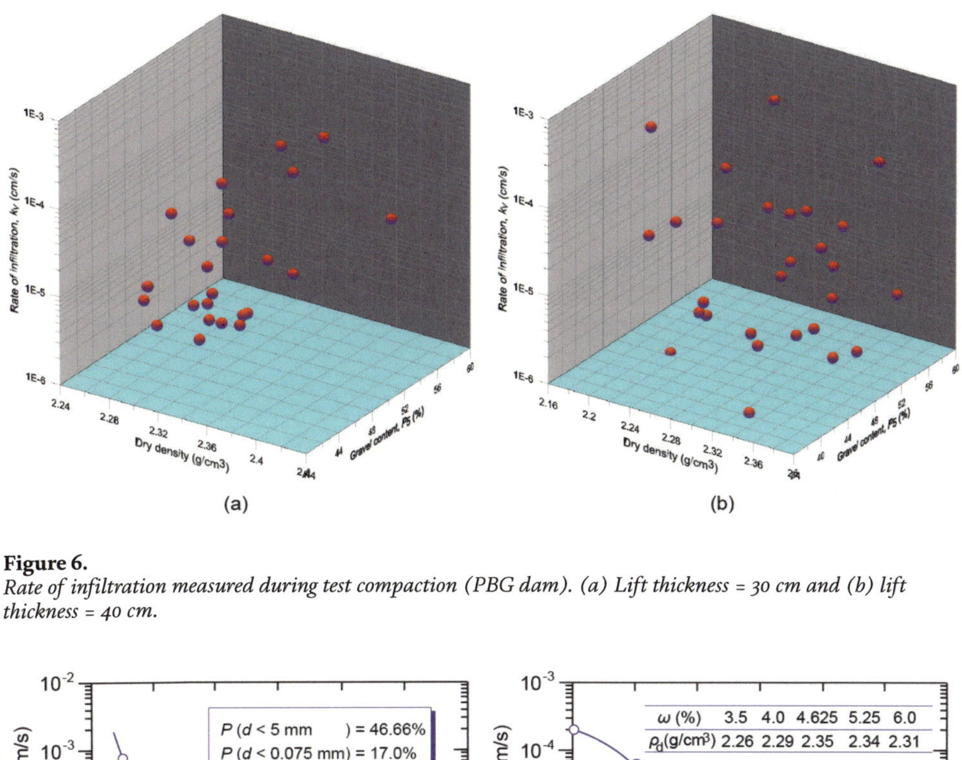

Figure 6.
Rate of infiltration measured during test compaction (PBG dam). (a) Lift thickness = 30 cm and (b) lift thickness = 40 cm.

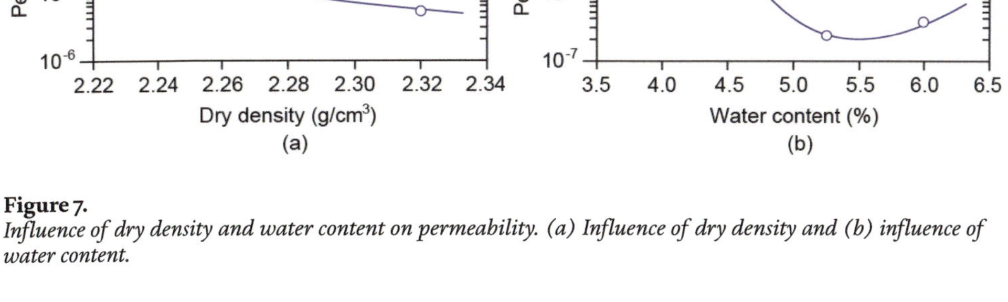

Figure 7.
Influence of dry density and water content on permeability. (a) Influence of dry density and (b) influence of water content.

wet of optimum results in a lower permeability although the resultant dry density is also lower than the maximum one (**Figure 7(b)**). It has been recognized very early that a dispersed microstructure (with a high degree of particle orientation) can be obtained for clay when it is compacted on the wet side (wetter than optimum) while compacting the same clay drier than optimum generally yields a flocculated microstructure (with a low degree of particle orientation) [17]. **Figure 8** replots the contours of permeability of a silty clay compacted by kneading action [18]. The lower permeability on the wet side than that on the dry side can be clearly observed from the distribution of contours.

Permeability is a good parameter indicating the pore structure. Since 1970s, the use of scanning electron microscopy and mercury intrusion porosimetry have shed sufficient light on the fabric of compacted clay [19]. It has been found that samples compacted dry of optimum tend to have a marked double-porosity fabric, with a macro-fabric consisting of large clay packets or macropeds separated by inter-packet voids as demonstrated in **Figure 9**. Samples compacted wet of optimum, on the other hand, have a relatively uniform fabric [20]. Therefore, dam CGM is better to be compacted slightly wetter than optimum. Another important benefit of compacting CGM wet of optimum is the less amount of wetting-induced collapse upon saturating during reservoir impounding [19, 20].

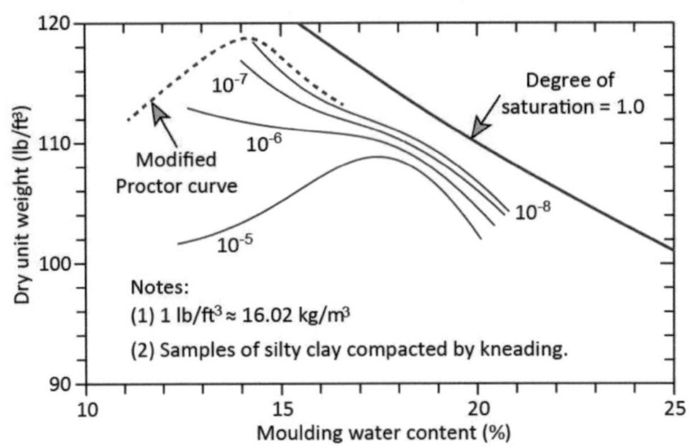

Figure 8.
Contours of permeability under saturated condition. From Mitchell et al. [18].

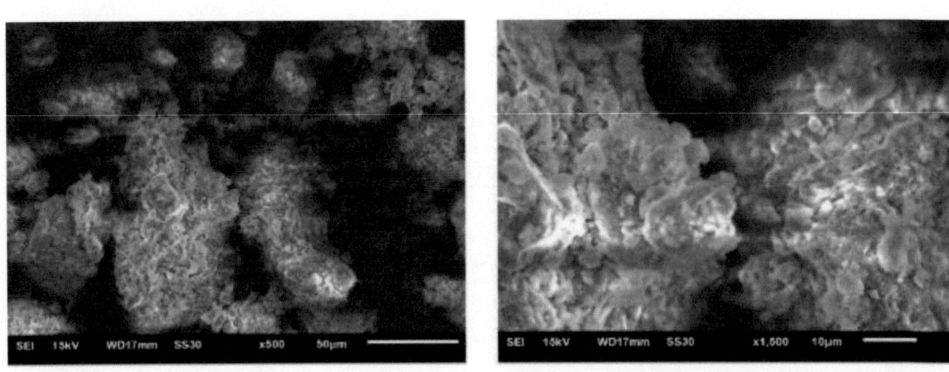

Figure 9.
Double-porosity fabric of the clay used in the LHK ECRD. (a) Amplified 500 fold and (b) amplified 1500 fold.

3.3 Strength

Two aspects of strength of CGM deserve attention in designing, i.e., the shear strength and the tensile strength. The shear strength is important for slope stability analyses and the tensile strength is useful in estimating the possibility of cracking under various conditions. The authors tested the CGM used in the CHB dam and investigated the influences of the gravel content on the strength components, i.e., the cohesion (c) and the friction angle (φ). The results are shown in **Figure 10**. Note each soil specimen was compacted to its maximum dry density under its optimum water content. The cohesion decreases steadily as a result of an increase in gravel content. On the contrary, the friction angle increases when the gravel content is increased. The CGM changes from a cohesive soil to a granular soil when the gravel content is gradually increased.

Uniaxial tensile experiments were also performed for the CHB CGM specimens with different gravel contents, and the typical results are shown in **Figure 11**.

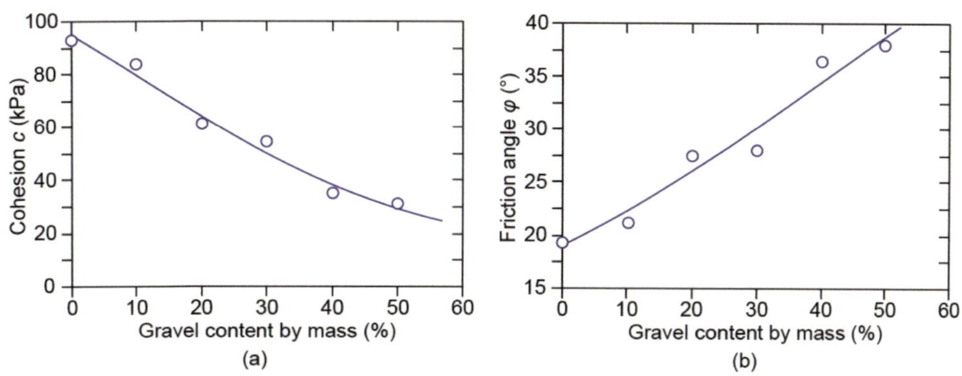

Figure 10.
Influence of gravel content on the shear strength. (a) Cohesion and (b) friction angle.

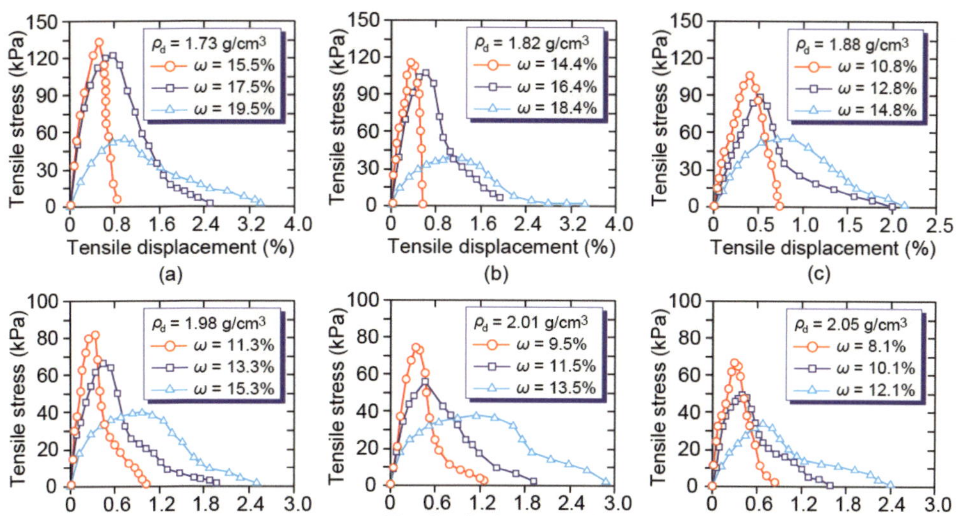

Figure 11.
Tensile strength of CGM with different gravel content. (a) $c_g = 0$*, (b)* $c_g = 10\%$*, (c)* $c_g = 20\%$*, (d)* $c_g = 30\%$*, (e)* $c_g = 40\%$*, (f)* $c_g = 50\%$*.*

For each gravel content, three different water contents were considered, i.e., ω_{opt}-2%, ω_{opt}, and ω_{opt}+2%. Variation of water content may occur as a result of sun exposure, rainfall or fluctuation of reservoir level. It can be seen in **Figure 11** that the tensile strength of CGM decreases significantly when the water content is increased. This is not surprise because matric suction of the unsaturated CGM contributes to its tensile strength and a wetter state indicates a lower matric suction. Despite of the higher strength of drier specimens, the tensile displacement corresponding to the peak tensile strength decreases when the water content is reduced, indicating an increasingly brittle response of specimens dry of optimum and increasingly ductile responses of specimens wet of optimum. **Figure 11** also shows that when the gravel content is increased from zero to 50% the tensile strength of specimens compacted to their densest states at their optimum water contents decreases gradually from 123 kPa to 50 kPa. These results indicate that CGM with a higher gravel content is more prone to crack, particularly when it is in a dry state.

3.4 Deformation

The deformation characteristics of CGM are generally studied by consolidated and drained triaxial compression experiments. **Figure 12** shows some typical stress vs. strain and volume change results obtained for CGMs used in the NZD and LHK dams. Strain-hardening behavior can be observed for both materials, particularly under a high confining pressure, e.g., when the confining pressure is over 2.0 MPa peak deviatoric stress cannot be attained even when the specimen is sheared to an axial strain of 15% as shown in **Figure 12(b)**. The volumetric response is generally contractive. However, shear dilation can present when the confining stress is low as illustrated in **Figure 12(a)**.

The general trends shown in **Figure 15** indicate that the deformation behavior of CGM can be well described by the constitutive model proposed by Duncan and Chang [21]. The model is based on a hyperbolic relationship between the deviatoric stress and the axial strain, and the tangential modulus (E_t) depends on the stress state including the minor principal stress (σ_3) and the stress level, i.e.

$$E_t = \left[1 - R_f \frac{(1-\sin\varphi)(\sigma_1 - \sigma_3)}{2c\cos\varphi + 2\sigma_3 \sin\varphi}\right]^2 \cdot k \cdot p_a \cdot \left(\frac{\sigma_3}{p_a}\right)^n \tag{15}$$

in which σ_1 denotes the major principal stress and p_a the atmospheric pressure. c and φ are shear strength parameters. R_f, k and n are three modulus parameters.

The axial strain and the radial strain can also be assumed following a hyperbolic relationship, which yields the final representation of the tangential Poisson ratio (ν_t) as follows [21]:

$$\nu_t = \frac{G - F\lg\left(\frac{\sigma_3}{p_a}\right)}{\left\{1 - \left[1 - R_f \frac{(1-\sin\varphi)(\sigma_1 - \sigma_3)}{2c\cos\varphi + 2\sigma_3\sin\varphi}\right]^{-1} \cdot \frac{D\cdot(\sigma_1 - \sigma_3)}{k\cdot p_a \cdot (\sigma_3 / p_a)^n}\right\}^2} \tag{16}$$

in which G, F, and D are another three parameters. **Table 2** summarizes the model parameters of the CGM used by different authors for the reviewed four dams. They may be used as references when similar cases are encountered.

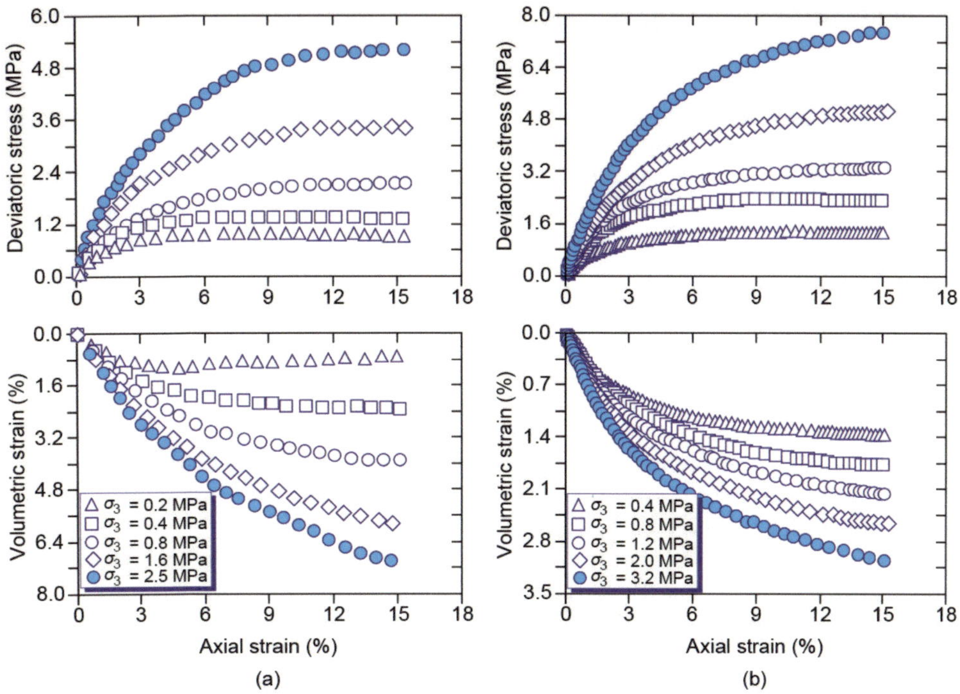

Figure 12.
Results of drained triaxial experiments on typical CGM. (a) CGM in the NZD dam and (b) CGM in the LHK dam.

Parameters	R_f	k	n	G	F	D	φ (°)	c (kPa)	ρ_d (g/cm^3)	Reference
PBG	0.78	550	0.31	0.31	0.07	4.0	34.0	60.0	2.36	[5]
	0.76	550	0.42	0.39	0.01	7.5	35.0	12.0	2.30	
CHB	0.87	646	0.40	0.40	0.04	2.0	31.1	12.0	2.22	[22]
	0.86	473	0.32	0.43	0.06	1.4	23.9	50.0	2.19	
NZD	0.63	405	0.47	0.31	0.04	5.9	33.0	60.0	1.93	[2]
	0.76	415	0.53	0.30	0.06	4.8	33.5	60.0	1.92	
LHK	0.72	650	0.36	0.36	0.02	3.0	31.0	40.0	2.16	[23]
	0.76	500	0.40	0.38	0.015	2.0	29.5	40.0	2.13	
	0.78	300	0.50	0.39	0.01	2.0	22.0	20.0	2.06	

Table 2.
Model parameters of the CGMs used in the reviewed dams.

The nonlinear elasticity model proposed by Duncan and Chang [21] is properly among the most widely used constitutive models because of its simplicity in using and convenience of parameter determination. However, some important features cannot be captured by this model. For example, the model is proposed based on triaxial compression tests and the influence of the intermediate principal stress is not considered. Furthermore, the model is a nonlinear elasticity one that the shear-induced volumetric contraction and dilation cannot be reflected. Many advanced constitutive models have been proposed for soils within the framework of different theories in the past decades [24–27]. However, they are beyond the scope of this chapter and are not perused further.

4. Construction techniques

Natural impervious materials excavated from borrow areas usually do not meet the specified gradation requirements. For instance, the raw materials obtained in the PBG and CHB dams contain much oversize particles, which need to be sieved out before placement. On the contrary, the raw materials in the NZD and LHK dams contains too much fine particles that the deformation moduli are not high enough and certain contents of coarse gravel should be added. Therefore, the most frequently required operations in preparing the core materials are screening and blending. This section gives some experiences used in the four dams reviewed.

4.1 Screening

Oversize particles can be removed at the borrow area before loading using hand labor or on the fill surface after dumping by using special rock rakes [4, 8]. Both techniques are not of sufficient efficiency, and have not been used for the dams reviewed. The grizzly is the commonly used particle-separating device in the PBG and CHB dams. A grizzly is a grating made of heavy bars, across which the material to be processed is passed. The bars are wider at the top than at the bottom, so that

Figure 13.
Screening system for the CHB dam (at the courtesy of Xue K). (a) Test screening device and (b) five screening stations.

the openings between bars increase in width with depth and, therefore, are not easy to clog by particles caught partway through [4]. The grizzly is often constructed with a sloping, vibrating grating, so that oversize particles are dumped over the end of the grizzly, while the desired material passes through. **Figure 13(a)** shows a test grizzly built for the CHB dam. The raw material was dumped from the truck on a high platform and fed into the vibrating grating (frequency = 500–850 r/min) through a slope trough inclined at 35° (from the horizontal plane). The bars were spaced 150 mm and had a slight inclination of 5°. The amplitude of vibration was 6–10 mm. The simple device shown in **Figure 13(a)** was used initially, and later five screening stations (**Figure 13(b)**) were built following similar concepts to produce the required large amount of materials. In the screening stations, the inclination of the vibrating grating was increased to 10°, and bars 3.5 m in length were used without welding.

The screening system for the PBG dam is shown in **Figure 14**. The raw material was first dumped onto a sloped grizzly (opening = 300 mm, size = 4.5 m × 6.0 m, and inclination = 30–34°) to remove the particles larger than 300 mm. The soils passing through the grizzly was then transferred by a belt conveyor to a specifically fabricated shaking screen with the purpose of sieving out the particles larger than 80 mm. The resultant material was transferred by a belt conveyor to the dam site directly. Belt conveyors are most suitable for moving large quantities of material over rough terrain where there are large differences in elevation between the dam site and the borrow pits or screening stations [4]. In the current case, a special tunnel 4 m in width and 3 m in height was excavated (lined with concrete), in which the belt conveyor was installed. The difference in elevation of the inlet and outlet is 460 m and the horizontal distance is 3985.84 m, indicating an average slope angle of 6.6°. The 1000 mm wide belt had a speed of 4 m/s and was capable of conveying 1000 ton of screened material per hour.

Trommel is another kind of screening device suggested by U.S. Bureau of Reclamation [4], which sieve out the oversize particles by rotating a cylinder of perforated sheet metal or wire screen. This device has not been practiced in the reviewed cases. Interested readers are referred to the relevant design standard [4].

Figure 14.
Screening system for the PBG dam (at the courtesy of Yao FH).

4.2 Blending

For the NZD and LHK dams, the low content of gravel particles in the original materials excavated from borrow areas necessitates the addition of gravel fraction in order to increase the deformation moduli of the cores. The common blending practice used in both NZD and LHK dams may be simply described as *spreading horizontally and excavating vertically* as shown in **Figure 15**. For the NZD dam, a layer of gravel 50 cm in thickness was first spread. Then a layer of natural clay 110 cm in thickness was spread on the filled gravel layer [28]. Three such interlayers were placed, forming artificial horizontal soil strata. Power shovels with a bucket volume of 6 m^3 were subsequently used to excavate the soil strata vertically from the bottom as shown in **Figure 15**. Sufficient mixing was achieved by running the open bucket through the clay-gravel mixture several times before loading. The thickness of the gravel and clay layers for the NZD dam was intended to adding 35% of artificial gravel particles to the natural soils. For the LHK dam, 40% of gravel particles were required to mix with natural soils, and this was achieved by placing a 50-cm thick gravel layer and an 83-cm thick clay layer sequentially. Shovels with a bucket volume of 4 m^3 were used to excavate the prepared strata as shown in **Figure 15(b)**.

4.3 Compaction

Padfoot rollers were used to compact the CGMs used in the four rockfill dams without exception. The selection of compacting unit, the number of passes, and the loose lift thickness were based on test filling. **Table 3** summarizes the compaction parameters for the CGM used in the four ECRDs. Some important points deserves

(a) (b)

Figure 15.
Blending techniques used in the NZD and LHK dams. (a) The NZD dam and (b) the LHK dam.

Dam	PBG	CHB	NZD	LHK
Loose lift thickness (cm)	45	30	27	30
Weight of compactor (ton)	25	26	22	26
Number of passes	8	2*+12	10	2*+10

*Note: *Number of passes of static compaction.*

Table 3.
Roller compaction parameters for the CGMs.

Figure 16.
Field compaction by GPS equipped padfoot rollers. (a) Scarification, (b) sprinkling, (c) compaction, and (d) quality checking.

to be mentioned. First, scarification of the surface of a compacted lift is always necessary prior to placing the next lift in order to ensure a good bond between the lifts. **Figure 16(a)** shows an example that scarification being performed by crawler tractors at the NZD dam. Second, the scarified loose surface layer may lose water due to evaporation, and watering the surface layer before placing the next lift is important. **Figure 16(b)** shows a truck sprinkling the surface of the LHK dam.

Strict compliance to the compaction specification is of great significance for a quality job. In the past two decades, global positioning system (GPS) technology has been increasingly used in dam engineering. GPS equipment has been carried by padfoot rollers in the NZD dam (**Figure 16(c)**) and later cases so that their routes of compaction and number of passes can be well monitored in the central control office. Once the action of a compaction roller deviates from the specification, remind or warn information can be send to the operator and immediate corrective measures can be taken. It is also important to note that blended soils generally have a wide grain size distribution. It is, therefore, necessary to perform quality checking more frequently to ensure that satisfactory results are obtained, as exemplified in **Figure 16(d)** taken from the LHK ECRD.

5. Field control test

Stability, deformation and seepage behaviors of ECRDs are generally evaluated using parameters obtained from laboratory and field tests where the samples are prepared at specified dry densities and water contents. This is the reason why soils must be placed as specified; otherwise, design assumptions may not be met and, in the worst case, unexpected distress might occur in the finished structure. The basic properties that should be checked frequently for the compacted soils are the dry density and water content. The standards for percent of compaction and water content are usually established for the total material [1, 2], and this necessitates a certain number of large-scale compaction tests if the prototype CGM is to be tested as the largest particles are up to 75–150 mm in diameter (**Table 1**). It is, however, generally not feasible to do this because testing total material could not keep pace with the rate of fill placement. Therefore, rapid field control test methods should be devised to guarantee the compaction quality. In this part, some methods used in the reviewed ECRDs are introduced.

5.1 Method used for the NZD dam

The main difficulty in evaluating the degree of compaction for CGM is the time required to obtain its compaction curve. For instance, 8 hours were required to carry out the three-point rapid compaction tests with a mold 300 mm in diameter [2]. However, if the fine fraction ($d < 20$ mm in the case of NZD dam) was tested with a mold 152.4 mm in diameter, the volume of materials to be tested and the compaction effort can be reduced considerably and the rapid compaction tests can be finished within only 1 h. Therefore, it is natural to establish a relationship between the percent compaction of the total material and that of the fine fraction. The percent compaction specified for the CGM in the NZD dam is 95% [2], and the corresponding percent compaction of the fine fraction was established as follows:

1. Perform the standard five-point compaction tests for the total material with a modified Proctor compaction mold (diameter = 600 mm and effort = 2690 kJ/m^3) and find out the maximum dry density (ρ_{dmax}) and optimum water content (ω_{opt}) of the total material.

2. For the maximum dry density sample, sieve out the particles with diameters over 20 mm. Weigh the coarse particles (m_g) and evaluate the total volume (V_g) of these particles using the density of the gravel particles (ρ_g).

3. Assume that the total material was compacted to 95% of the maximum dry density, the total volume (V) of the sample can be evaluated as $V = m/(0.95\rho_{dmax})$ where m is the dry mass of the total material. The dry density of the fine fraction can then be evaluated by $\rho_c = (m-m_g)/(V-V_g)$.

4. Perform the standard five-point compaction tests for the fine fraction ($d < 20$ mm) with a standard Proctor compaction mold (diameter = 152.4 mm and effort = 595 kJ/m^3), and find out the maximum dry density (ρ^*_{dmax}) and optimum water content (ω^*_{opt}) of the fine fraction.

5. The percent compaction of the fine fraction can be calculated by ρ_c/ρ^*_{dmax}.

6. Repeat the above procedures for other contents of added gravel materials so that a reasonable range of percent compaction can be obtained for the fine fraction ($d < 20$ mm).

Figure 17 shows such a relationship established parallel by two companies. It has been found that the 95% of compaction for the total material (2690 kJ/m^3) can always be achieved when the fine fraction is compacted to 98% of its maximum dry density (595 kJ/m^3). Therefore, this criterion was used in field to check the quality of compaction for the NZD dam. Only the dry density of the fine fraction needs to be calculated and the three-point rapid compaction tests were performed with the fine fraction.

The percent of compaction of the total material can be evaluated conveniently once its maximum dry density is known. Although this is time consuming and not practical in field, the maximum dry density of the total material can be evaluated a priori. For the NZD dam, the maximum dry densities of the total material with various contents of added gravel particles were obtained and plotted against P_{20} in **Figure 18**. Once the P_{20} after compaction is known, the maximum dry density and thus the percent compaction of the total material can be estimated using **Figure 18**. Both the percent compaction of the total material and that of the fine fraction were checked during the construction of the NZD dam. Back analyses using the pore water pressures measured by piezometers show that the permeability coefficient of the core material is of the magnitude of 10^{-9} cm/s, which is considerably lower than the values measured both in laboratory and in field [29].

5.2 Method used for the PBG dam

The maximum diameter of CGM in the PBG dam is 80 mm. Compaction criteria were specified for the fine fraction only (d < 5 mm), i.e., $\omega^* = \omega^*_{opt}$ + (1–2)% and

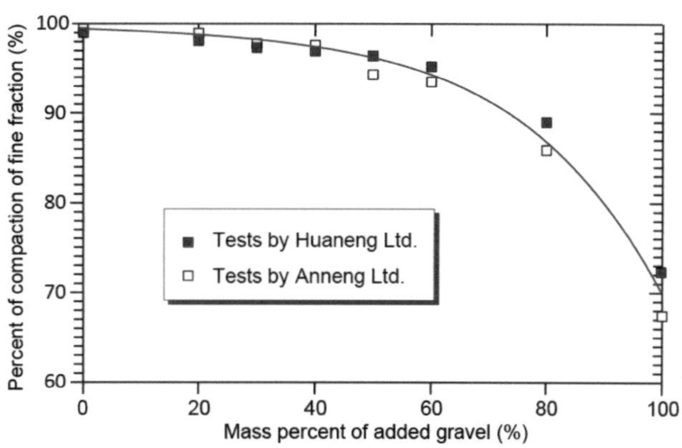

Figure 17.
Percent of compaction of the fine fraction corresponding to 95% compaction of the total material.

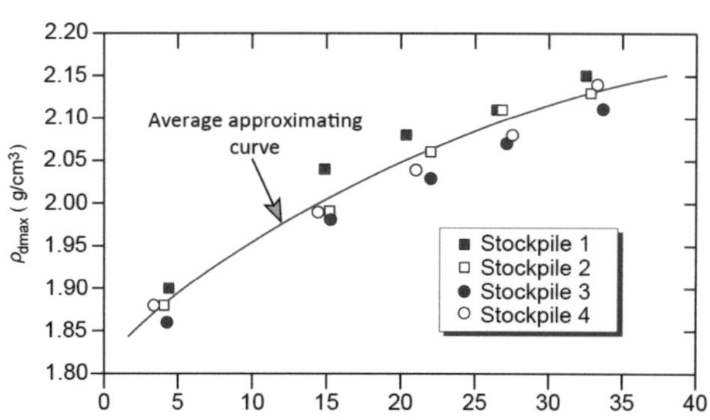

Figure 18.
Maximum dry density of total material with different P_{20}.

$\rho^*_d/\rho^*_{dmax} \geq 98\%$ (2740 kJ/m^3) or 100% (595 kJ/m^3). It is logical to perform three-point rapid compaction tests on fine fraction to check the criteria. However, an even simpler strategy was also used for the PBG dam. The procedure is described as follows. For a given dry density of the fine fraction (ρ^*_d), the relationship between the dry density of the total material (ρ_d) and the gravel content (c_g) is similar to Eq. (4) and Eq. (9), i.e.

$$\rho_d = \frac{\rho_g}{\left(1-c_g\right)\rho_g + c_g\rho_d^*}\rho_d^* \tag{17}$$

which is illustrated by the red curve in **Figure 19**. If the field water content of the total material (ω) is obtained, the wet density of the total material (ρ) at this water content can be estimated for each gravel content as illustrated by the blue curve in **Figure 19**. This blue curve is the border distinguishing the unacceptable samples from those acceptable ones. For instance, the sample denoted by point A has a wet density above the blue curve. This is realistic only when the dry density of the fine fraction is higher than ρ^*_d and thus the sample is acceptable. On the contrary, sample B has a wet density below the curve, indicating that its fine fraction has not been compacted to the dry density of ρ^*_d and therefore is unacceptable.

5.3 Method used for the CHB dam

Water content and gravel content are also important indices to be checked for the total material. For instance, determining the gravel content is the prerequisite of using the concepts given in **Figures 17** and **18**. However, evaluate the mass of water contained in the prototype material needs to dry the tested total material up to 8 h, and this is obviously not acceptable for a rapid check test. Bao et al. suggested a practical way to determine the water content and gravel content for the CHB dam [6, 7]. First, it is easy to verify that the water content of the total material (ω) can be calculated by the water content of the fine fraction (ω^*) and that of the gravel fraction (ω_g), that is

$$\omega = \left(1-c_g\right)\omega^* + c_g\omega_g \tag{18}$$

in which c_g is again the mass percent of gravel particles (dry materials).

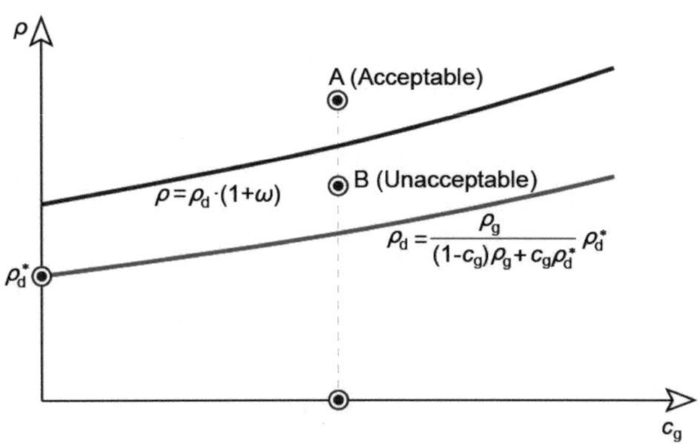

Figure 19.
Relationship between the wet density and gravel content.

Now assume that a test pit has been excavated and the total volume (V) of the pit has been obtained by filling water or sand. The total wet mass of the excavated materials (M) can also be obtained immediately. The total material is then sieved into two parts, i.e., the gravel particles ($d \geq 5$ mm) and the fine fraction ($d < 5$ mm). The sample of fine fraction can be dried quickly by open alcohol flame, and the water content (ω^*) is easily obtained. Meanwhile, the sieved gravel particles are washed with clean water and wiped with dry towels. The wet mass of these clean gravel particles (M_g) are weighed as soon as possible. The wet mass of the fine fraction (M_c) can now be determined as $M_c = M - M_g$. Consequently, the dry mass of the fine fraction (m_c) is obtained, i.e.

$$m_c = \frac{M_c}{1+\omega^*} \tag{19}$$

The key assumption adopted by Bao et al. is that all the gravel particles are in a saturated state with dry surfaces. This water content is normally in the range of 1.5–3.5%, with an average of 2.3% in the CHB case. With this average value for ω_g, the dry mass of the gravel particles (m_g) can be readily obtained, i.e.

$$m_g = \frac{M_g}{1+\omega_g} \tag{20}$$

The mass content of the gravel content is calculated by $c_g = m_g/(m_g + m_c)$, and the water content of the total material can be estimated by Eq. (18). The volume of the gravel particles (V_g) can either be measured directly or be calculated using its wet mass and apparent wet density [6, 7]. Afterward, the volume of fine fraction (V_c) is obtained as $V_c = V - V_g$. Then, the dry density of the total material (ρ_d) and the fine fraction (ρ^*_d) can be obtained as follows:

$$\rho_d = \frac{m_c + m_g}{V}, \quad \rho_d^* = \frac{m_c}{V_c} \tag{21}$$

This rapid method can be used to assess the compaction quality for both the total material and the fine fraction. However, two aspects need to be pointed out. First, the fine fraction is dried by alcohol flame, which can result in inaccuracy in water content because the high temperature applied to the soil can drive off the adsorbed water and burn or drive off volatile organic matter, neither of which should be removed in a normal water content test [30]. Second, the water content assumed for the gravel particles may influence the results to an unacceptable level. Preliminary check tests should be performed before using. **Figure 20** compares the water contents of total material samples obtained by alcohol flaming and normal oven drying as well as the gravel contents obtained by two methods. The difference in water content by the two methods varies in the range of −0.6–0.6%, with an average of −0.1%, indicating the reliability of the flaming method for this particular case. Furthermore, the gravel contents obtained by the two methods are also very close to each other with a slight difference ranging between −0.4 and 0.3%.

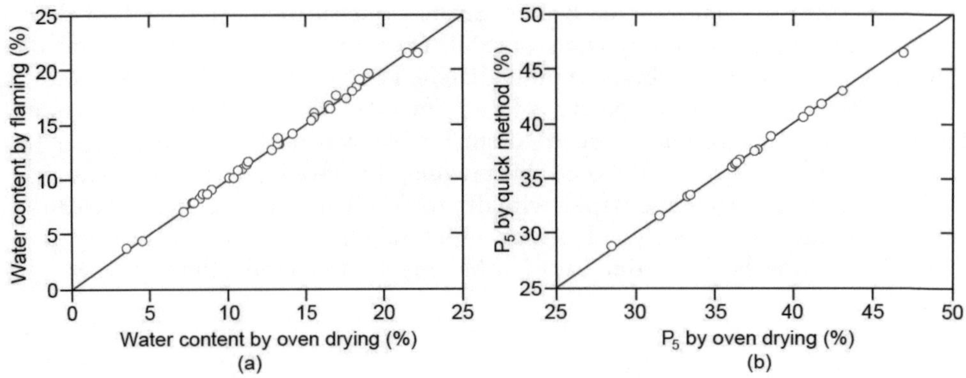

Figure 20.
Verification of the method used in the CHB dam. (a) Water content verification and (b) gravel content verification.

Two aspects are concerned as can be summarized from the above cases, i.e., the degree of compaction of the total material and that of the fine fraction (the diameter defining the fine fraction varies from dam to dam, e.g. 20 mm in the NZD dam and 5 mm in the CHB dam). The percent compaction of the total material is useful to guarantee the overall deformation performance but is not enough to ensure the impermeability of the fine fraction. On the other hand, the percent compaction of the fine fraction is a good indication of the impermeability but is not enough for the overall behavior of the total material. The concept of quality control for high ECRDs is increasingly stringent. For the PBG dam, emphasis was placed on the fine fraction. For the NZD dam, three-point rapid compaction tests were performed to check the compaction quality of the fine fraction ($d < 20$ mm). The percent compaction of the total material was not checked every point but was checked periodically. For the CHB dam, both the percent of compaction of the total material and the fine fraction were checked simultaneously, and this concept is used similarly in the LHK ECRD under construction. Deformation and seepage behaviors of these dams monitored in field prove the effectiveness of these control test methods.

6. Conclusions

Clay and gravel mixture has been used in constructing impermeable system of embankment dams for many years. The compaction performance as well as the strength, deformation and permeability behaviors are considerably influenced by the quantity of gravel contained. Determining and controlling the gravel content is therefore of great significance in design and construction. Many compaction tests performed for the reviewed cases in this chapter show that the fine fraction can be compacted to its densest state when the gravel content is below about 30%, beyond which the maximum dry density of the total material continues to increase while the percent compaction of the fine fraction decreases, as evidenced by the abrupt increase in permeability. A peak for the maximum dry density of the total material can generally be achieved when the gravel content is around 70%, beyond which the dry density decreases considerably when the gravel content is further increased.

Removing oversize particles from a cohesive soil is generally not easy, especially when the soil is in a wet state. However, experience obtained previously shows the success of using slope grizzlies and some shaking /vibrating screens or both. When a certain amount of gravel is to be blended with a raw material, the spreading and excavation practice seems to be an effective way. Scarification, sprinkling, and field checking should be performed to ensure the compaction and bonding quality and confirm the design assumptions. It is now a trend that both the percent compaction of the fine fraction and that of the total material be verified using some rapid field control tests as exemplified in this chapter.

Properties that have not been discussed but are also very important to the safety of embankment dams include the dynamic behavior, wetting-induced collapse behavior, and creep behavior of clay and gravel mixtures. Laboratory tests have been performed and various constitutive models that describing these important behaviors have been proposed and incorporated into finite element procedures, which play important role in predicting the performance of dams to be built. Field instrumentations for settlement, earth pressure and pore water pressure have also provided valuable information on the safety status of constructed dams. All these advances have contributed to the successful construction and operation of the reviewed cases and will continue to play important roles in even higher earth and rockfill dams.

Acknowledgements

This work is supported by the National Natural Science Foundation of China (Nos. 51779152 & U1765203).

Author details

Zhongzhi Fu*, Shengshui Chen, Enyue Ji, Guoying Li and Yang Lu
Geotechnical Engineering Department, Nanjing Hydraulic Research Institute, Nanjing, China

*Address all correspondence to: fu_zhongzhi@yahoo.com

References

[1] China Ministry of Water Resources. Design Specification for Rolled Earth-Rock Fill Dams (SL 274-2001). Beijing: China Water & Power Press; 2011

[2] Zhang ZL. Key Technologies for Earth Core Rockfill Dams Higher Than 200 m and Their Engineering Application. Beijing: China Water & Power Press; 2011. ISBN: 978-7-5084-4447-5

[3] USBR (U.S. Bureau of Reclamation). Chapter 2: Embankment design. In: Design Standards No. 13 Embankment Dams. Denver; 2012

[4] USBR (U.S. Bureau of Reclamation). Chapter 10: Embankment construction. In: Design Standards No. 13 Embankment Dams. Denver; 2012

[5] Yao FH, Yang XG. Key Technologies in PuBuGou Gravely Earth Core Rockfill Dam. Beijing: China Water & Power Press; 2015. ISBN: 978-7-5170-3293-9

[6] Bao HF, Wang HB, Pang G, Wang K. Research on rapid detection method for filling quality of impervious core wall with gravelly soil in dam. Journal of Water Resources and Architectural Engineering. 2014;**12**(1):105-111. DOI: 10.3969/j.issn.1672-1144.2014.01.022

[7] Bao HF, Pang G, Yang YB, Hu P. Study on the compaction characteristics and maximum dry density of gravel soil. Geotechnical Investigation and Surveying. 2016;**9**:22-28

[8] Xue K, Liu XS, Yao KW. Techniques used in removing oversize particles in the gravely soils used in the ChangHeBa hydropower station. Sichuan Water Power. 2015;**34**(3):25-30. DOI: 10.3969/j.issn.1001-2184.2015.03.006

[9] Zhang ZL, Feng YL, Xiang B, Yuan YR. Design, researches and practices of impervious materials for core wall of Nuozhadu Hydropower Station. Chinese Journal of Geotechnical Engineering. 2013;**35**(7):1323-1327

[10] Xiong L, Yang JP. Sensitivity analysis of the mixing parameters for the clay gravel mixtures used in the LiangHeKou dam and the controlling measures. Construction of Water Conservancy and Hydropower Engineering. 2017;**5**:62-66

[11] Che WB, Yang JP, Jiang WH, Chen X. Analysis on roller compaction tests of gravel soil material in LiangHeKou hydropower station. Water Power. 2018;**44**(2):28-32. DOI: 10.3969/j.issn.0559-9342.2018.02.008

[12] Lu Y. Experimental study and mechanism analysis of engineering properties of clay mixed with gravel under cold environments [thesis]. Nanjing: Hohai University; 2019

[13] ASTM. Standard Test Methods for Laboratory Compaction Characteristics of Soil Using Standard Effort (D 698-00). West Conshohocken; 2000

[14] ASTM. Standard Test Methods for Laboratory Compaction Characteristics of Soil Using Modified Effort (D 1557-02). West Conshohocken; 2002

[15] Shelley TL, Daniel DE. Effect of gravel on hydraulic conductivity of compacted soil liners. Journal of Geotechnical Engineering. 1993;**119**(1):54-68

[16] ASTM. Standard Test Method for Infiltration Rate of Soils in Field Using Double-Ring Infiltrometer (D 3385-03). West Conshohocken; 2003

[17] Lambe TW. The engineering behaviour of compacted clay. Journal of Soil Mechanics and Foundations Division. 1958;**84**(2):1-35

[18] Mitchell JK, Hooper DR, Campanella RG. Permeability of compacted clay. Journal of Soil Mechanics and Foundations Division. 1965;**91**(4):41-65

[19] Alonso EE, Pinyol NM, Gens A. Compacted soil behavior: Initial state, structure and constitutive modelling. Géotechnique. 2013;**63**(6):463-478. DOI: 10.1680/geot.11.P.134

[20] Sivakumar V, Wheeler SJ. Influence of compaction procedure on the mechanical behaviour of an unsaturated compacted clay. Part 1: Wetting and isotropic compression. Géotechnique. 2000;**50**(4):359-368

[21] Duncan JM, Chang CY. Nonlinear analysis of stress and strain in soils. Journal of Soil Mechanics and Foundations Division, ASCE. 1970;**96**(5):1629-1653

[22] Xiong K, He YL, Wu XY, Dong YL. Stress and deformation behavior of foundation gallery of ChangHeBa hydropower station. Chinese Journal of Geotechnical Engineering. 2011;**33**(11):1767-1774

[23] Chen ZB, Zhu JG. Three-dimensional finite element analysis on stress-strain and materials parameters sensibility of LiangHeKou core rockfill dam. Journal of Fuzhou University (Natural Science). 2010;**38**(6):893-899

[24] Huang MS, Yao YP, Yin ZY, Liu EL, Lei HY. An overview on elementary mechanical behaviors, constitutive modeling and failure criterion of soils. In: Proceedings of the 12th National Conference on Soil Mechanics and Geotechnical Engineering; Engineering. Shanghai; 17-21 July 2015. pp. 1-26

[25] Kolymbas D. An outline of hypoplasticity. Archive of Applied Mechanics. 1991;**61**:143-151

[26] Zienkiewicz OC, Chan AHC, Pastor M, Schrefler BA, Shiomi T. Computational Geomechanics with Special Reference to Earthquake Engineering. New York: John Wiley & Sons; 1999

[27] Fu ZZ, Chen SS, Wei KM. A generalized plasticity model for the stress-strain and creep behavior of rockfill materials. Science China Technological Sciences. 2019;**62**:1-19. DOI: 10.1007/s11431-018-9362-3

[28] Wang YF, Chi X. Techniques for blending clay-gravel mixtures for 300-m extremely high rockfill dams. Construction and Maintenance of Water Project. 2018;**9**:5-10. DOI: 10.16616/j.cnki.11-4446/TV.2018.09.02

[29] Wu YK, Zhang BY, Yu YZ, Zhang ZL. Consolidation analysis of NuoZhaDu high earth-rockfill dam based on the coupling of seepage and stress-deformation-physical state. International Journal of Geomechanics. 2016;**16**(3):1-11. Article No: 04015085. DOI: 10.1061/(ASCE)GM.1943-5622.0000555

[30] USACE (U.S. Army Corps of Engineers). Construction control for earth and rock-fill dams. In: Engineering Manual. EM 1110-2-1911. Washington, DC; 1995

Chapter 2

Heightening of an Existing Embankment Dam: Results from Numerical Simulations

Zhu Yumeng, Guoying Li, Zhankuan Mi, Zhongzhi Fu and Kuangmin Wei

Abstract

The old dam of the Zhushou Reservoir is a clay core rock-debris dam with a maximum height of 63.4 m. After heightening, the new dam is a concrete-faced rockfill dam with a maximum height of 98.1 m. In the initial design stage, a rigid connection is proposed between the cutoff wall and toe slab. After the concrete cutoff wall is built at the axis of the old dam, a complete cutoff system is composed of cutoff wall, toe slab, and face slab. In this paper, based on the static and dynamic tests of dam materials, the Shen Zhujiang double-yield surface elastic-plastic model is used as the static constitutive model, and the contact friction model is used as the contact surface model. The three-dimensional finite element method is used to simulate the construction filling and water storage process during operation. The simulation results show that the maximum horizontal displacement occurs in the dam body of the old dam and the maximum settlement occurs at the interface between the old and new dams. During the storage period, the cutoff wall will not be damaged, and the tensile stress of the local area at the junction of toe slab and bank slope has exceeded the allowable value for C30 plain concrete, so the reinforcement should be strengthened at this location.

Keywords: Zhushou Reservoir, heightening, core rockfill dam, face rockfill dam, impervious system

1. Introduction

With the rapid development of the economy and the large-scale development of water energy, the construction of reservoir dams has become an important engineering initiative to meet the needs of social and economic development. Over time, the sediment in the reservoir continues to accumulate, and the storage capacity for prosperity and flood control continues to decrease or even loses capacity completely. In addition, due to a lack of water level data and drainage area data in the original design or a lack of labor, equipment, funds, or other resources during construction, the construction of small storage capacity reservoirs cannot meet the current demand for water resources. Therefore, the construction of new water conservancy facilities or the heightening of the old dams has become an urgent problem to consider. Compared with the construction of a new dam, raising an original dam body does not require the consideration of the location of a new dam, and it can obtain a larger storage capacity at a lower economic cost. Therefore, the dam heightening scheme has attached increasing attentions from engineers [1].

Addressing the technical problems that rise during the process of heightening is becoming a top priority due to the large amount of work and the complexity of construction technology. There are different key problems in dam heightening engineering due to the dam type and heightening method. Earth-rock dams are a widely used type of dam. Due to the permeability of earth-rock materials, it is urgent to study the impact of seepage on the earth-rock dam during the heightening process [2]. For slope-type heightened and thickened concrete dams, the key issues related to dam heightening are the stress concentrations and deformation of the dam body during construction and operation, stress analysis and structural form of the interface between new and old concrete, and design of drainage and water stop [3]. Periodic changes in the temperature and changes in the temperature of the old dam after new concrete is placed will cause problems such as deterioration of the dam heel stress, cracks in the joint surface, and cracks in the downstream dam surface [4].

There are many engineering precedents for dam elevation, such as the Goscheneralp Dam and Grande Dixence Dam in Switzerland, Steenbras Dam in South Africa, Roseires Dam in Sudan, and Danjiangkou Dam and Songyue Dam in China [3, 5–9]. Due to the rapid increases in the urban population of Cape Town, raising the Steenbras Dam offered an effective solution to the problem of a serious water shortage. During the course of anchoring the dam, engineers considered that post-stressing would have advantages in terms of cost and expedition. Essentially the process is one of placing vertical cables through the wall of a mass concrete dam from the crest into the foundation and stressing the cables to produce stabilizing compressive forces on the upstream face. Similar to the Steenbras Dam, the Songyue Dam also raised the dam to meet the water supply needs of Helong City. The Songyue Dam is located in a severely cold area, with an average annual temperature of 4.8°C, and the temperature changes greatly during the year. Therefore, the heightened structure needed to adapt to the characteristics of the severely cold area. The calculation research on the Songyue Dam heightening scheme shows that setting a sliding joint in the middle of the joint surface can absorb the shrinkage and deformation of a part of the newly poured concrete, which has a significant effect on improving the tensile stress of the upstream and downstream dam surfaces.

The Zhushou Reservoir is located in Sichuan Province, China, which is located in a seismically active area. The dam of Zhushou Reservoir is a clay core rock-debris dam. To meet the production and domestic water demand of nearby cities, it is necessary to expand the capacity of the Zhushou Reservoir. Under the action of gravity loads, water loads, and earthquake loads, effectively coordinating the deformation of the rockfill of the new and old dams to allow the stress and deformation of the seepage control system to be within the allowable range of the materials is a major technical difficulty to be solved. Therefore, based on the experience of previous engineering technologies, the necessary theoretical research is carried out to accurately predict the stress and deformation of the dam, especially the coordination between the deformation of the old and new dams, to improve the rationality of engineering design and to improve future engineering operations.

2. Project description

2.1 General situation of the Zhushou Reservoir project

The Zhushou Reservoir pivotal project is located in Liangshan Prefecture, Sichuan Province, and is a medium-sized reservoir. The dam is made of a clay core rock-debris dam. Its top elevation is 2416.10 m, the dam length is 190.00 m, the top elevation of the wave-proof wall is 2417.10 m, the dam height is 63.4 m, and the width of the dam top is 6.0 m. Both the upper and lower dams are provided with rockfilled prisms. The upstream slope is protected by dry block stone, while the downstream slope is protected by a dry block stone arch ring and turf in the circle. The thickness of the dry block stone is 40 cm. The top width of the gravel soil core wall is 6.0 m, the top elevation is 2415.3 m, and the upper and lower slopes are 1:0.4.

2.2 Dam heightening scheme

According to the water supply project planning of the Baihetan hydropower station resettlement area, to meet the production and domestic water demand of the resettlement area, the Zhushou Reservoir should be expanded and matched to the corresponding water diversion project. The dam should be increased from 63.4 m to 98.1 m. At the same time, when the dam is heightened, the impervious body of the original dam should be strengthened [10].

The objective of dam heightening is to make use of the water-retaining capacity of the original core wall dam to produce rockfill heightening on the top and downstream slope of the old dam so that the original dam body becomes a part of the heightened dam. At the same time, a core wall and foundation anti-seepage system of the original dam is strengthened, a concrete cutoff wall is added, and the foundation anti-seepage curtain grouting is strengthened. The anti-seepage type of the heightening dam body adopts the upstream reinforced concrete-faced slab, the slope ratio of the upstream dam is 1:1.4, and the comprehensive slope of the downstream rockfill body is 1:1.6 [11]. **Figures 1** and **2** show general view of the Zhushou Reservoir dam.

2.3 Dam heightening construction procedure

To avoid excessive deformation and cracking of the lower core wall caused by the compression of the upper high rockfill, the cutoff wall is constructed after the upper rockfill body is filled and settled for 3 months. The concrete connecting plate between the cutoff wall and the toe slab shall be constructed after the toe slab and the face plate are completed.

The overall construction procedure is as follows: old dam filling → new dam filling to 2447.90 m → core wall reinforcement and cutoff wall construction → toe slab

Figure 1.
Plane figure of heightening of the Zhushou Reservoir dam.

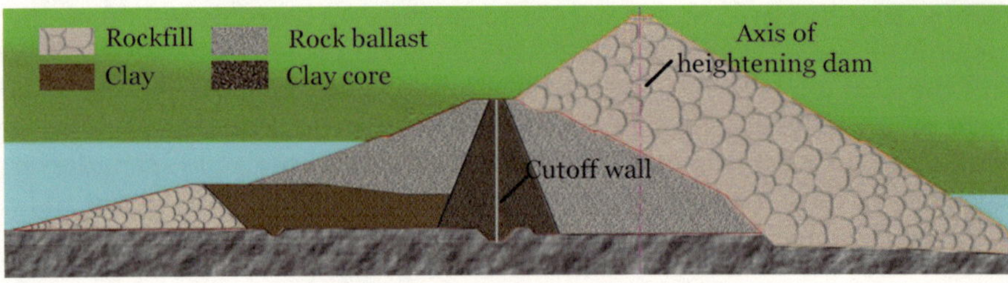

Figure 2.
Standard profile of heightening of the Zhushou Reservoir dam.

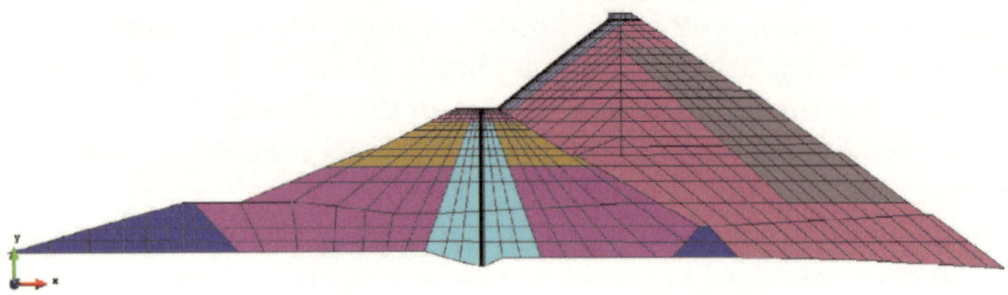

Figure 3.
Finite element mesh diagram of a typical riverbed section.

construction → panel construction → connecting plate construction → new dam filling to 2451 m. The water level remains at 2395.0 m during the construction period. The construction period of dam heightening is 31 months, which are as follows:

From September of the first year to February of the second year, the construction period of the old dam filling is 6 months.

From March of the second year to November of the second year, the construction period of the new dam filling to an elevation of 2447.9 m is 9 months.

From December of the second year to May of the third year, the construction period of core wall reinforcement and cutoff wall construction is 6 months.

During June of the third year, the construction period of toe slab is 1 month.

From July of the third year to August of the third year, the construction period of concrete panel and wave wall construction is 2 months.

From September of the third year to November of the third year, the construction period of connecting plate construction is 3 months.

From December of the third year to July of the fourth year, the construction period of new dam filling to 2451 m is 4 months.

3. Finite element modeling

3.1 Finite element meshing

Figure 3 shows a finite element mesh diagram of a typical riverbed section, **Figure 4** shows a three-dimensional finite element mesh diagram, and **Figure 5** shows an anti-seepage system (core wall, connecting plate, toe slab and panel) meshing diagram, where the X forward direction is defined as from the left bank to the right bank, the Y forward direction is defined as upstream to downstream, and the Z forward direction is defined as the opposite direction of gravity. The three-dimensional solid element adopts an 8-node hexahedral isoparametric element and a 4-node tetrahedral isoparametric element, and the latter is treated as a degenerated hexahedral element. There are 29,905 generating units and 33,482 nodes in total.

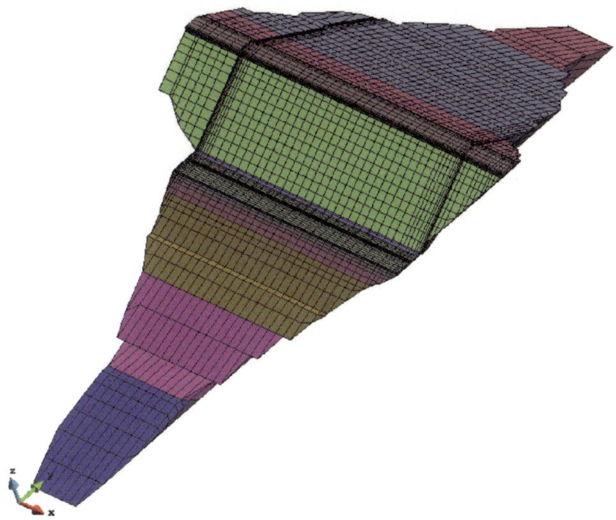

Figure 4.
Three-dimensional finite element mesh diagram.

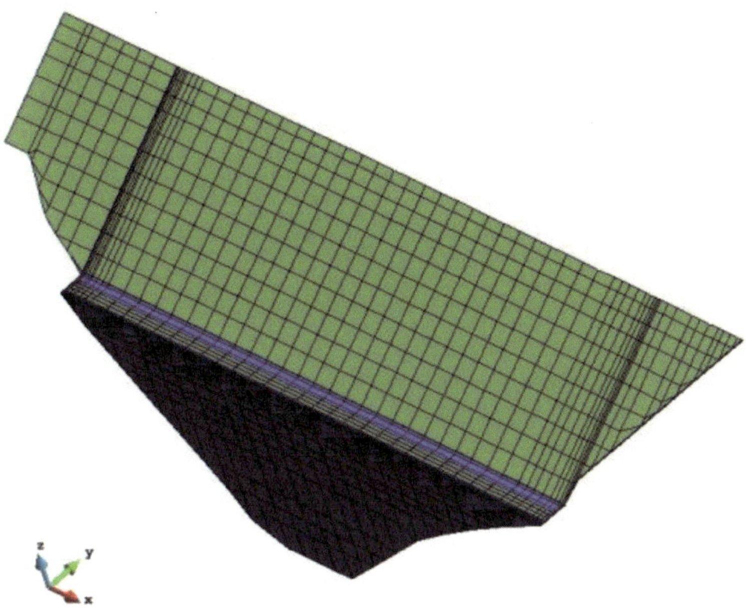

Figure 5.
Anti-seepage system meshing diagram.

3.2 Construction sequence and simulation of the water storage process

According to the above construction and water storage process, the order of the filling and storage simulation in the finite element calculation is as follows: old dam filling → new dam filling to 2447.90 m (the water level remained at 2395.0 m) → cutoff wall construction → toe slab construction → panel construction → connecting plate construction → new dam filling to 2451 m → upstream water storage to a normal water level elevation of 2444 m. There are 70 stages for simulation, including 42 stages for dam filling and 38 stages for water storage. **Figure 6** shows the simulation diagram of the Zhushou Reservoir construction and

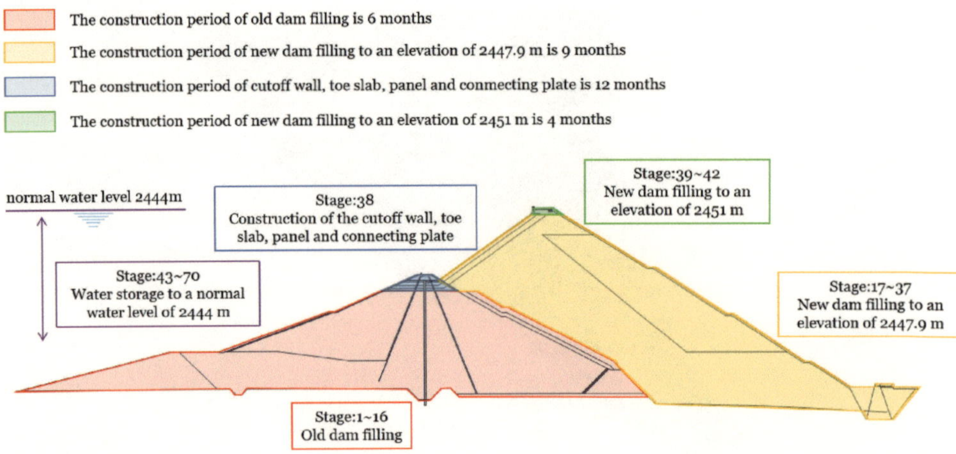

Figure 6.
The Zhushou reservoir construction and water storage process simulation diagram.

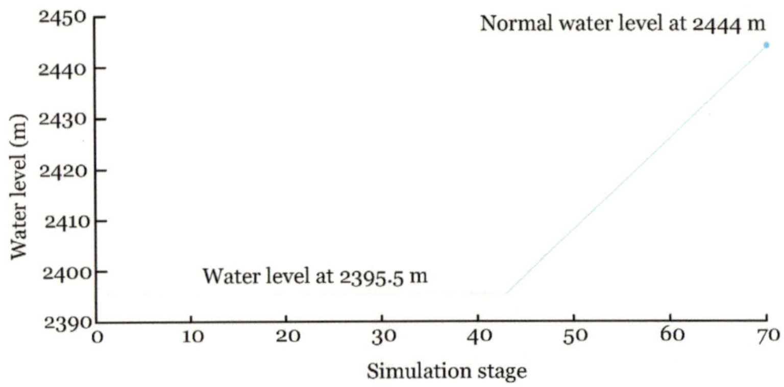

Figure 7.
The Zhushou reservoir water level-time curve during the construction and water storage process.

water storage process. **Figure** 7 shows the water level-time curve of the Zhushou Reservoir during the construction and water storage process.

4. Constitutive model of the dam material

4.1 Constitutive model of the soil and rockfill

As the main body of the concrete-faced rockfill dam, reasonable simulation of its stress–strain relationship is very important to improve the rationality of the calculation results of the stress and deformation of the concrete-faced rockfill dam. In this project, the constitutive model of rockfill material is based on the Shen Zhujiang double-yield surface elastic-plastic model proposed by Shen Zhujiang. Compared with the nonlinear elastic model, the model can consider the dilatancy and shear-shrinkage characteristics of rockfill bodies and can more accurately reflect the stress-strain characteristics of dam bodies than other models.

In the Shen Zhujiang double-yield surface elastic-plastic model, the two-yield surfaces are only regarded as the boundary of elastic region and are no longer related to hardening parameters. The double-yield surface is used to establish the unloading criterion, make the elastic-plastic matrix symmetrical, and specify the direction of plastic strain. As shown in **Figure 8**, due to the double-yield surface, not only the loading direction B will produce plastic strain, but also the loading directions A and C will produce plastic strain.

The two-yield surfaces of the Shen Zhujiang double-yield surface elastic-plastic model are

$$\left.\begin{array}{l} F_1 = p^2 + r^2q^2 - f_1 = 0 \\ F_2 = q^s/p - f_2 = 0 \end{array}\right\} \tag{1}$$

where $p = \frac{1}{3}(\sigma_1 + \sigma_2 + \sigma_3)$, $q = \frac{1}{\sqrt{2}}\left[(\sigma_1 - \sigma_2)^2 + (\sigma_2 - \sigma_3)^2 + (\sigma_3 - \sigma_1)^2\right]^{1/2}$, and r and s are model parameters and can be taken as 2 for rockfill materials. The expression of the strain increment of the double-yield surface model is as follows:

$$\{\Delta\varepsilon\} = [D]^{-1}\{\Delta\sigma\} + A_1\{n_1\}\left\{\frac{\partial f_1}{\partial\sigma}\right\}^T\{\Delta\sigma\} + A_2\{n_2\}\left\{\frac{\partial f_2}{\partial\sigma}\right\}^T\{\Delta\sigma\} \tag{2}$$

where $[D]$ is the elastic matrix, $\{n_1\}$ and $\{n_2\}$ are the normal direction cosines of the yield surface, and A_1 and A_2 are plasticity coefficients. Δf_1 and Δf_2 can be written as.

$$\left.\begin{array}{l} \Delta f_1 = 2p\Delta p + 2r^2q\Delta q \\ \Delta f_2 = q^s/p(-\Delta p/p + s\Delta q/q) \end{array}\right\} \tag{3}$$

The model adopts the normal flow rule, so the plastic potential surface is orthogonal to the direction of the plastic strain increase and $Q_1 = F_1$ and $Q_2 = F_2$. According to the normal flow rule, $\partial Q_1/\partial\sigma_1$, $\partial Q_1/\partial\sigma_3$, $\partial Q_3/\partial\sigma_1$, and $\partial Q_3/\partial\sigma_3$ can be calculated. Considering that $P = \frac{1}{3}(\sigma_1 + 2\sigma_3)$, $q = \sigma_1 - \sigma_3$, and $\Delta\varepsilon_v = \Delta\varepsilon_1 + 2\Delta\varepsilon_3$ under triaxial conditions, Eq. (2) can be expressed as:

$$\left.\begin{array}{l} \dfrac{\Delta\varepsilon_1}{\Delta\sigma_1} = \dfrac{1}{E} + \dfrac{4}{9}\left(p + 3r^2q\right)^2 A_1 + \dfrac{1}{9}\left(\dfrac{1}{p} - \dfrac{3s}{q}\right)^2 \dfrac{q^{2s}}{p^2} A_2 \\ \dfrac{\Delta\varepsilon_v}{\Delta\sigma_1} = \dfrac{1-2\nu}{E} + \dfrac{4}{3}p\left(p + 3r^2q\right)A_1 + \dfrac{1}{3}\left(\dfrac{1}{p} - \dfrac{3s}{q}\right)\dfrac{q^{2s}}{p^2} A_2 \end{array}\right\} \tag{4}$$

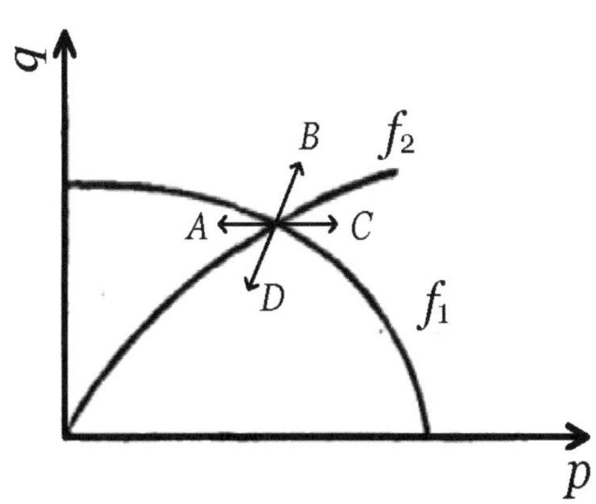

Figure 8.
Double hardening model.

Tangent Young's modulus is defined as $E_t = \frac{\Delta\sigma_1}{\Delta\varepsilon_1}$, and tangent volume ratio is defined as $\mu_t = \frac{\Delta\varepsilon_v}{\Delta\varepsilon_1}$. According to E_t and μ_t, A_1 and A_2 can be expressed as

$$\left.\begin{aligned} A_1 &= \frac{1}{4p^2}\frac{\eta\left(\frac{9}{E_t}-\frac{3\mu_t}{E_t}-\frac{3}{G_e}\right)+2s\left(\frac{3\mu_t}{E_t}-\frac{1}{B_e}\right)}{2(1+3r^2\eta)(s+r^2\eta^2)} \\ A_2 &= \frac{p^2q^2}{q^{2s}}\frac{\left(\frac{9}{E_t}-\frac{3\mu_t}{E_t}-\frac{3}{G_e}\right)-2r^2\eta\left(\frac{3\mu_t}{E_t}-\frac{1}{B_e}\right)}{2(3s-\eta)(s+r^2\eta^2)} \end{aligned}\right\} \quad (5)$$

where $\eta = q/p$ and G_e and B_e are the elastic shear modulus and bulk modulus, respectively:

$$G_e = E_{ur}/2(1+\nu) \quad (6)$$

$$B_e = E_{ur}/3(1-2\nu) \quad (7)$$

In the formula, the elastic Poisson's ratio ν is 0.3, and E_{ur} is the modulus of the unloading resilience. The tangent Young's modulus E_t and tangent volume ratio μ_t in Eq. (5) are two basic variables of the model, which are expressed as follows:

$$E_t = KP_a\left(\frac{\sigma_3}{P_a}\right)^n(1-R_fS_l)^2 \quad (8)$$

$$\mu_t = 2c_d\left(\frac{\sigma_3}{P_a}\right)^{n_d}\frac{E_iR_s}{\sigma_1-\sigma_3}\frac{1-R_d}{R_d}\left(1-\frac{R_s}{1-R_s}\cdot\frac{1-R_d}{R_d}\right) \quad (9)$$

$$S_l = \frac{(\sigma_1-\sigma_3)(1-\sin\phi)}{2c\cos\phi+2\sigma_3\sin\phi} \quad (10)$$

where P_a is the atmospheric pressure, K is the Young's modulus coefficient, and n is the power of the tangent Young's modulus E_t, which increases with the increase in the confining pressure σ_3. R_f is the failure ratio, S_l is the stress level, and c and ϕ are the shear strength indexes; $R_s = R_fS_l$. R_d, C_d, and n_d are calculation parameters; C_d corresponds to the maximum shrinkage volume strain when σ_3 equals the unit atmospheric pressure; n_d is the power of the shrinkage volume strain which increases with the increase in σ_3, and R_d is the ratio of $(\sigma_1-\sigma_3)_d$ to the asymptotic value of the deviating stress $(\sigma_1-\sigma_3)_{ult}$ when the maximum shrinkage occurs. The elastoplastic matrix of the double-yield surface model can be obtained from the inverse of Eq. (2):

$$\{\Delta\sigma\} = [D]_{ep}\{\Delta\varepsilon\} \quad (11)$$

However, the expression of $[D]_{ep}$ is quite complex. In order to simplify the $[D]_{ep}$, Prandtl-Reuss flow rule is adopted in π plane; $[D]_{ep}$ can be expressed as

$$[D]_{ep} = \begin{bmatrix} d_{11} & & & & & \\ d_{21} & d_{22} & & & \text{symmetric} & \\ d_{31} & d_{32} & d_{33} & & & \\ d_{41} & d_{42} & d_{43} & d_{44} & & \\ d_{51} & d_{52} & d_{53} & d_{54} & d_{55} & \\ d_{61} & d_{62} & d_{63} & d_{64} & d_{65} & d_{66} \end{bmatrix} \quad (12)$$

where $d_{11} = M_1 - P(S_x + S_x)/q - QS_x^2/q^2$, $d_{22} = M_1 - P(S_y + S_y)/q - QS_y^2/q^2$, $d_{33} = M_1 - P(S_z + S_z)/q - QS_z^2/q^2$, $d_{44} = G_e - Q\tau_{xy}^2/q^2$, $d_{55} = G_e - Q\tau_{yz}^2/q^2$, $d_{66} = G_e - Q\tau_{zx}^2/q^2$, $d_{12} = M_2 - P(S_x + S_y)/q - QS_xS_y/q^2$, $d_{13} = M_2 - P(S_x + S_z)/q - QS_xS_z/q^2$, $d_{14} = -P\tau_{xy}/q - QS_x\tau_{xy}/q^2$, $d_{15} = -P\tau_{yz}/q - QS_x\tau_{yz}/q^2$, $d_{16} = -P\tau_{zx}/q - QS_x\tau_{zx}/q^2$, $d_{23} = M_2 - P(S_y + S_z)/q - QS_yS_z/q^2$, $d_{24} = -P\tau_{xy}/q - QS_y\tau_{xy}/q^2$, $d_{25} = -P\tau_{yz}/q - QS_y\tau_{yz}/q^2$, $d_{26} = -P\tau_{zx}/q - QS_y\tau_{zx}/q^2$, $d_{34} = -P\tau_{xy}/q - QS_z\tau_{xy}/q^2$, $d_{35} = -P\tau_{yz}/q - QS_z\tau_{yz}/q^2$, $d_{36} = -P\tau_{zx}/q - QS_z\tau_{zx}/q^2$, $d_{45} = -P\tau_{yz}/q - Q\tau_{xy}\tau_{yz}/q^2$, $d_{46} = -P\tau_{zx}/q - Q\tau_{xy}\tau_{zx}/q^2$, and $d_{56} = -P\tau_{zx}/q - Q\tau_{yz}\tau_{zx}/q^2$. In the formula, $S_x = \sigma_x - p$, $S_y = \sigma_y - p$, $S_z = \sigma_z - p$, $M_1 = K_p + 4G_e/3$, $M_2 = K_p - 2G_e/3$, $P = B_eG_e\gamma/(1 + B_e\alpha + G_e\delta)$, and $Q = G_e^2\delta/(1 + B_e\alpha + G_e\delta)$ are as follows:

$$K_p = \frac{B_e}{1 + B_e\alpha}\left(1 + \frac{B_eG_e\gamma^2}{1 + B_e\alpha + G_e\delta}\right) \tag{13}$$

$$\alpha = \frac{A_1}{r^2} + \eta^2A_2 \tag{14}$$

$$\beta = r^2\eta^2A_1 + s^2A_2 \tag{15}$$

$$\gamma = \eta(A_1 - sA_2) \tag{16}$$

$$\delta = \beta + B_e\left(\alpha\beta - \gamma^2\right) \tag{17}$$

The Shen Zhujiang elastic-plastic model has eight model parameters, which are $K, n, R_f, c, \phi, R_d, C_d$, and n_d. The parameters can also be determined from the results of conventional triaxial tests. Compared with the parameters of the Duncan $E - \nu$ model, only the latter three parameters R_d, C_d, and n_d of the Shen Zhujiang elastic-plastic model are different from those of the Duncan model.

The Shen Zhujiang elastic-plastic model can also be calculated by the parameters of the model Duncan $E - \nu$ model. Its tangent volume ratio μ_t can be obtained from the tangent Poisson's ratio ν_t:

$$\mu_t = 1 - 2\nu_t \tag{18}$$

The tangent Poisson's ratio ν_t in the Duncan $E - \nu$ model is as follows:

$$\nu_t = \frac{G_e - F \cdot \lg(\sigma_3/Pa)}{\left[1 - D(\sigma_1 - \sigma_3)/E_i\left(1 - R_fS_l\right)\right]^2} \tag{19}$$

For unloading, the modulus of resilience is calculated as follows:

$$E_{ur} = K_{ur}P_a\left(\frac{\sigma_3}{P_a}\right)^n \tag{20}$$

where K_{ur} is the modulus of resilience.

The loading and unloading criteria of the Shen Zhujiang elastic-plastic model are as follows:

a. If $F_1 > F_{1\max}$ and $F_2 > F_{2\max}$, it is fully loaded. $A_1 > 0$ and $A_2 > 0$.

b. If $F_1 > F_{1\max}$ and $F_2 \le F_{2\max}$ and $F_1 \le F_{1\max}$ and $F_2 > F_{2\max}$, it is partially unloaded. $A_1 = 0$ or $A_2 = 0$.

c. If $F_1 \leq F_{1\max}$ and $F_2 \leq F_{2\max}$, it is fully unloaded. $A_1 = A_2 = 0$.

For coarse-grained materials, $c = 0$ and ϕ is calculated by the following formula:

$$\phi = \phi_o - \Delta\phi \lg\left(\frac{\sigma_3}{P_a}\right) \tag{21}$$

where ϕ_0 and $\Delta\phi$ are material parameters determined by triaxial test results.

4.2 Constitutive model of concrete

The linear elastic model is used for concrete materials, and the stress–strain relationship conforms to the following generalized Hooke's law:

$$\{\sigma\} = [D]\{\varepsilon\} \tag{22}$$

where $[D]$ is an elastic matrix.

4.3 Interface model

At present, the Goodman thickness-free elements and Desai thin-layer elements are commonly used. Because the interface is a kind of interface without a thickness, it is more suitable to use the Goodman element without a thickness to theoretically simulate the interface. However, in fact, a Goodman element without a thickness must obtain a large normal stiffness to avoid overlap. In addition, shear dislocation does not necessarily occur on the interface and may penetrate into the soil at a certain distance. Desai thin-layer elements reflect normal deformation to a certain extent, but the choice of the thickness of thin-layer elements has a great influence on the calculation results. A large element thickness will introduce errors in the physics, and a small element thickness will introduce errors in the mathematics. Desai suggests that the ratio of the thickness t to the length B is as follows:

$$\frac{t}{B} = \frac{1}{10} \sim \frac{1}{100} \tag{23}$$

For the constitutive model of the contact surface, the hyperbolic model and ideal elastic-plastic model of the relationship between the shear stress and relative displacement proposed by Clough and Duncan are most commonly used. The results show that the shear stress on the interface between the soil and structure is not uniform, the shear deformation is actually a rigid-plastic deformation, and the contact friction model can be well simulated.

Before the shear stress τ on the contact surface reaches the destructive shear stress τ_f, the dislocation deformation is very small, and the displacement of the contact surface is mainly caused by shear deformation. When the shear stress τ reaches the destructive shear stress τ_f, the displacement of the contact surface is mainly caused by the dislocation deformation and can develop indefinitely.

The deformation on the contact surface can be divided into two parts: basic deformation and failure deformation. The basic deformation is similar to the deformation calculation model of other soils, expressed as $\{\Delta\varepsilon'\}$. The failure deformation includes sliding failure and tension cracking failure, which only exists when the shear stress of the element reaches the shear strength and sliding failure along the contact surface or tension cracking failure occurs on the contact surface. In $\{\Delta\varepsilon''\}$, the total deformation of the contact surface is as follows:

$$\{\Delta\varepsilon\} = \{\Delta\varepsilon'\} + \{\Delta\varepsilon''\} = [C']\{\Delta\sigma\} + [C'']\{\Delta\sigma\} \tag{24}$$

There are two forms of failure and deformation of the elements: tension cracking and slip. The rigid-plastic model is used to calculate the relative shear deformation of the element. There is no relative slip on the contact surface before failure, and after failure, the relative slip will continue to develop.

For the three-dimensional thin-layer contact surface element, the Y direction is the normal direction of the contact surface:

$$\begin{Bmatrix} \Delta\varepsilon_x \\ \Delta\varepsilon_y \\ \Delta\varepsilon_z \\ \Delta\gamma_{xy} \\ \Delta\gamma_{yz} \\ \Delta\gamma_{zx} \end{Bmatrix} = \begin{bmatrix} C_{11} & C_{12} & C_{13} & 0 & 0 & 0 \\ C_{21} & C_{22}+\dfrac{1}{E'} & C_{23} & 0 & 0 & 0 \\ C_{31} & C_{32} & C_{33} & 0 & 0 & 0 \\ 0 & 0 & 0 & C_{44}+\dfrac{1}{G_{xy}{}'} & 0 & 0 \\ 0 & 0 & 0 & 0 & C_{55}+\dfrac{1}{G_{yz}{}'} & 0 \\ 0 & 0 & 0 & 0 & 0 & C_{66} \end{bmatrix} \begin{Bmatrix} \Delta\sigma_x \\ \Delta\sigma_y \\ \Delta\sigma_z \\ \Delta\tau_{xy} \\ \Delta\tau_{yz} \\ \Delta\tau_{zx} \end{Bmatrix} \tag{25}$$

If the contact surface is under tension, E' can be set to a very small value, such as E' = 5.0 kPa. If the contact surface is under compression, a larger value should be taken, or $\frac{1}{E'} = 0.0$.

The value of G' is determined by the stress level of the contact surface element: when the stress level is $S < 0.99$, the value of G' is larger, or $\frac{1}{G'} = 0.0$; when the stress level is $S > 0.99$, the shear failure of the contact surface element occurs, and the shear modulus corresponding to the residual stiffness of the element, or G' = 5.0 kPa, is obtained.

The flexibility matrix $[C]$ of the contact surface is directional. After forming the stiffness matrix $[C]^{-1}$ in the local coordinate system, the element stiffness matrix in the global coordinate system needs to be obtained by a coordinate transformation.

4.4 Material parameters

4.4.1 Concrete

The dam uses C25 concrete and C30 concrete. The unit weight, elastic modulus and Poisson's ratio of C25 and C30 concrete are 2.5 t/m^3, 2.8×10^4 N/mm^2, and 0.167 and 2.5 t/m^3, 3.0×10^4 N/mm^2, and 0.167, respectively.

4.4.2 Dam material

The lithology of the newly filled rockfill material in the Zhushou Reservoir is the Ordovician Hongshiya Formation (O1h) quartz sandstone, fine sandstone with silty mudstone, and Qiaojia Formation (O2q) gray thin-to-medium thick sandstone, dolomite, and limestone. According to design filling standards and field testing data, the triaxial CD test had been carried out on rockfill materials of the heightening dam and the filling materials of the old dam body. The calculated parameters determined by the test are shown in **Table 1**.

	Material name	ρ_d (g/cm^3)	c (kPa)	φ_o (°)	$\Delta\varphi$ (°)	k	n	R_f	D	F	G
Old dam	Gravel clay core wall	1.84	56.9	29.3	0	164.4	0.46	0.69	3.68	0.1	0.3
	Stone slag in the upper part of the dam hell (elevation above 2390 m)	2.04	0	41.8	9.1	318.8	0.46	0.79	2.78	0.04	0.35
	Stone slag in the lower part of the dam shell (elevation above 2390 m)	2.12	0	44.9	9.1	431.3	0.38	0.72	3.3	0.09	0.37
	Rockfill	2.14	0	47.7	10.1	811	0.31	0.54	10.4	0.12	0.4
New dam	Cushion zone	2.2	0	58.8	10.9	1245.6	0.35	0.60	10.4	0.12	0.4
	Transition region	2.17	0	59.5	13.3	1405.4	0.29	0.65	10.1	0.15	0.39
	Main rockfill area	2.16	0	59.4	13.6	1301.5	0.27	0.60	9.3	0.15	0.39
	Secondary rockfill area	2.14	0	56.1	11.9	954.1	0.37	0.63	9.5	0.13	0.36

Table 1.
Calculated parameters of the dam material.

5. Results and discussion

5.1 Stress and deformation of the dam body

Considering the stress and deformation of the new dam after filling and storage period and influence of the stress and deformation of the new dam on the old dam, **Table 2** lists the characteristic values of the stress and deformation of the dam body.

Figures 9 and **10** show the contour of the displacements of the dam body during the completion period and the storage period. The simulation results show that the maximum horizontal displacement occurs in the dam body of the old dam and the maximum settlement occurs at the interface between the old and new dams. During the completion period, the maximum settlement of the dam is 47.5 cm, and the horizontal displacement to the upstream and downstream is 18.2 cm and 6.90 cm, respectively. After the water storage, the maximum deformation of the dam body under upstream water load was reduced to 10.2 cm, while the horizontal displacement towards the downstream was increased to 9.25 cm, and the maximum settlement was increased to 48.8 cm.

Statistical items		Dam body	
		Completion period	Storage period
Displacement along the river (cm)	Downstream	−18.2	−10.2
	Upstream	6.9	9.25
Settlement (cm)		47.5	48.8
Major principal stress (MPa)		2.13	2.14
Minor principal stress (MPa)		1.21	1.23

Table 2.
Characteristic values of stress and deformation of the dam body.

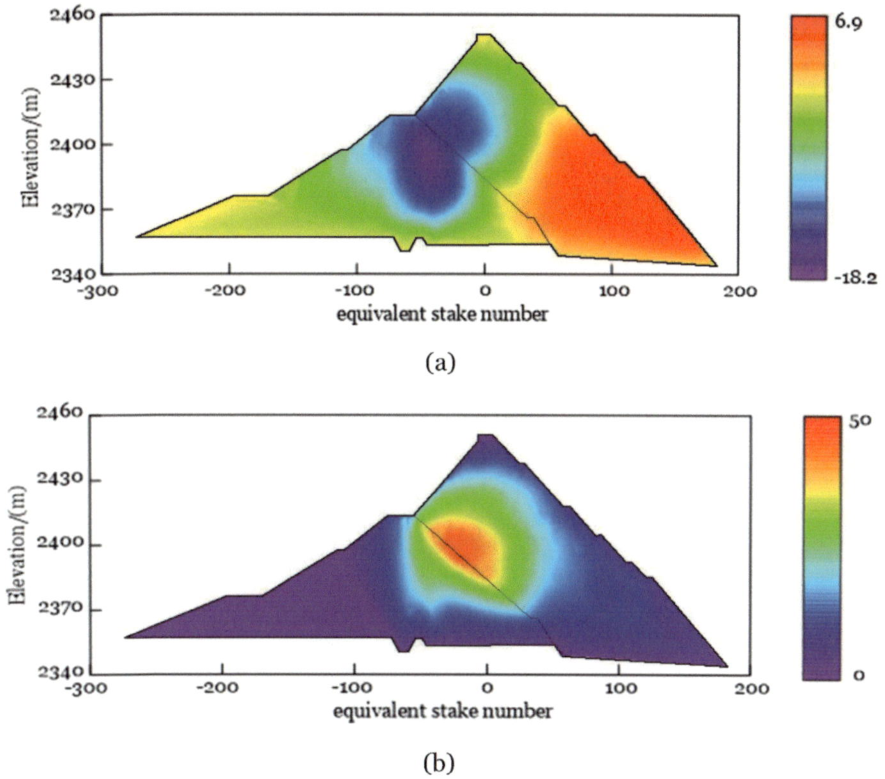

Figure 9.
Contour of the displacements of the dam body during the completion period (cm). (a) Displacement along the river and (b) settlement.

The results of principal stress calculation show that due to the large modulus of cutoff wall and pile foundation, significant stress concentration has occurred in the dam.

5.2 Stress and deformation of the cutoff wall

Table 3 lists the characteristic values of the stress and deformation of the cutoff wall.

Since the cutoff wall is constructed after the new dam is filled to 2447.9 m, the deformation of the cutoff wall will not occur during the completion period, so only the deformation distribution during the storage period is given. **Figure 11** shows contour of the displacement of the cutoff wall during the storage period. The axial displacement of the dam is represented by the compression from both sides towards the riverbed, and the deformation in the direction of the right bank and the left bank is 0.12 cm and 0.11 cm, respectively. The axial displacement of the dam is generally small. For the displacement along the river, the water load shows a deformation towards the downstream, and the maximum value is 10.6 cm. Because the upper part of the impervious wall is filled with rockfill and supported by the connecting plate, the deformation along the river of the impervious wall increases first and then decreases slightly from the bottom to the top. For the vertical displacement, the maximum value is 0.48 cm, which increases gradually from the bottom to top under the action of the upper water load.

Figure 12 shows the contour of the dam axial stresses on the downstream and upstream sides of the cutoff wall during the completion period. **Figure 13** shows the

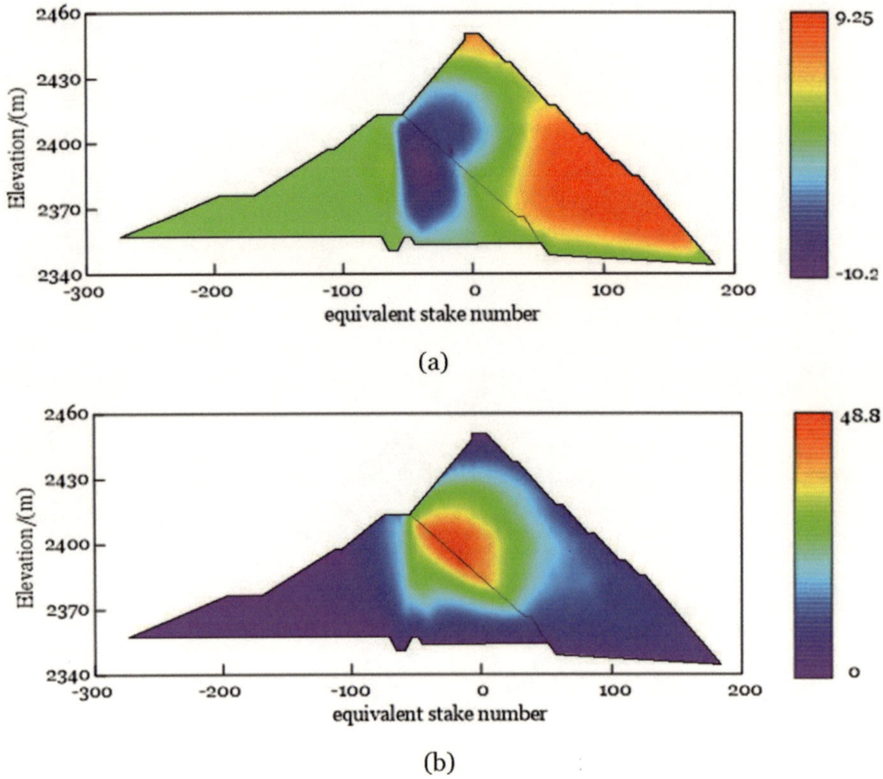

Figure 10.
Contour of the displacements of the dam body during the storage period (cm). (a) Displacement along the river and (b) settlement.

Statistical items		Cutoff wall	
		Completion period	Storage period
Dam axial displacement (cm)	Left side bank	/	−0.11
	Right side bank	/	0.12
Displacement along the river (cm)	Downstream	/	10.6
Settlement (cm)		/	0.48
Dam axial stress (MPa)	Tensile stress	−0.21	−2.53
	Compressive stress	1.18	3.21
Major principal compressive stress (MPa)		2.25	12.0
Minor principal tensile stress (MPa)		−0.23	−1.74

Table 3.
Characteristic values of stress and deformation of the cutoff wall.

contour of the dam axial stresses on the downstream and upstream sides of the cutoff wall during the storage period. Because the cutoff wall will be built after the new dam is basically completed, the stress difference between the upstream and downstream faces of the completion period is small, the stress of the cutoff wall is mainly caused by the self-weight, and the tensile and compressive stresses are small. During the storage period, the axial stress of the dam corresponds to the deformation direction. After storage, the upstream face is in tension at both ends of the middle compression zone, while the downstream face is basically in compression,

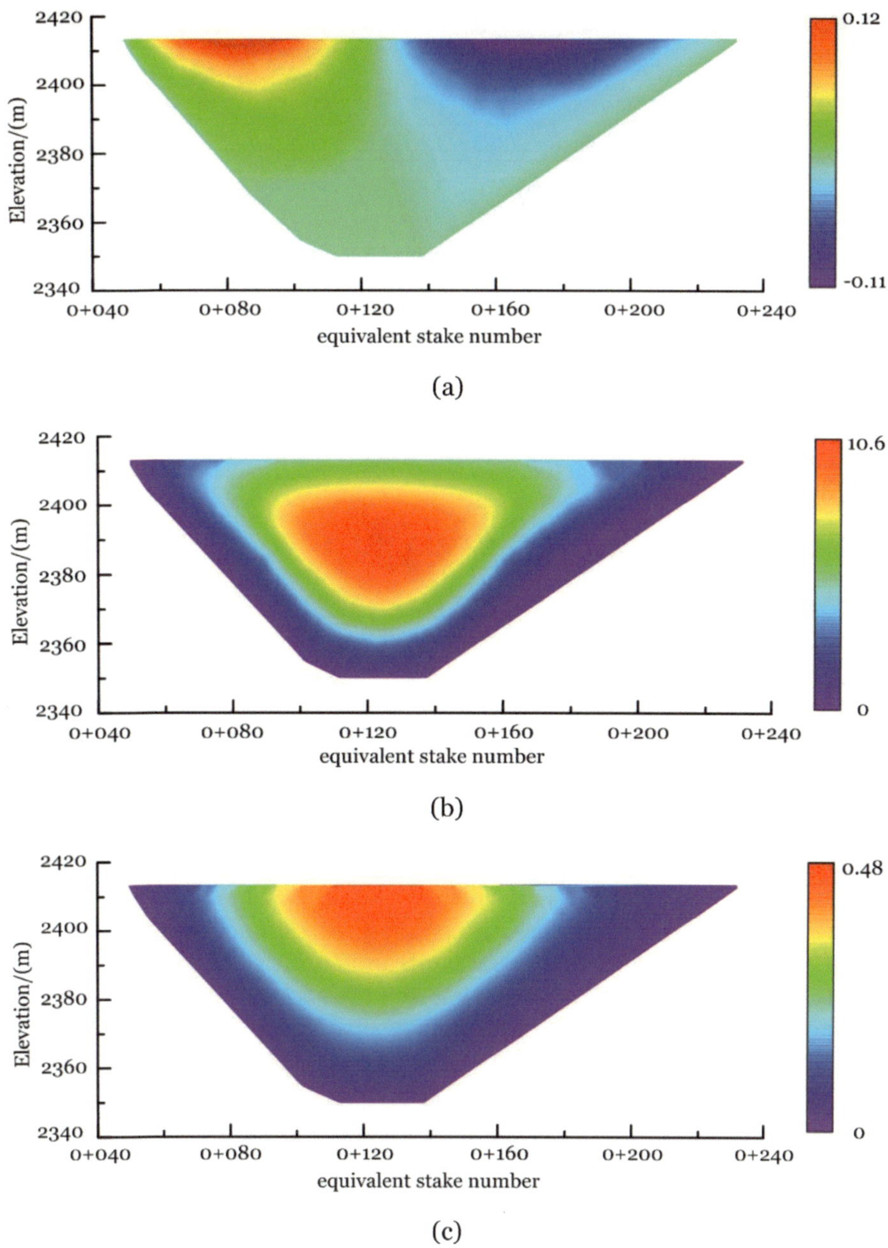

Figure 11.
Contour of the displacements of the cutoff wall during the storage period (cm). (a) Dam axial direction, (b) displacement along the river and (c) settlement.

but the pressure stress at both sides is significantly greater than that at the riverbed. The maximum value of the tensile and compressive stress is −2.53 MPa and 3.21 MPa, respectively. For the major principal stress, the downstream stress is greater than the upstream stress because the deformation is oriented downstream during the storage period. At the same time, due to the relatively small height of the wall near the bank slope and the influence of the boundary constraints, the local stress near the bank slope is concentrated, so the stress at the bank slope on both banks is large, and the maximum pressure stress is 12.0 MPa. For the minor principal stress, the upstream and downstream faces are all in compression at the middle part and tension at both sides. The maximum tensile stress is −1.74 MPa.

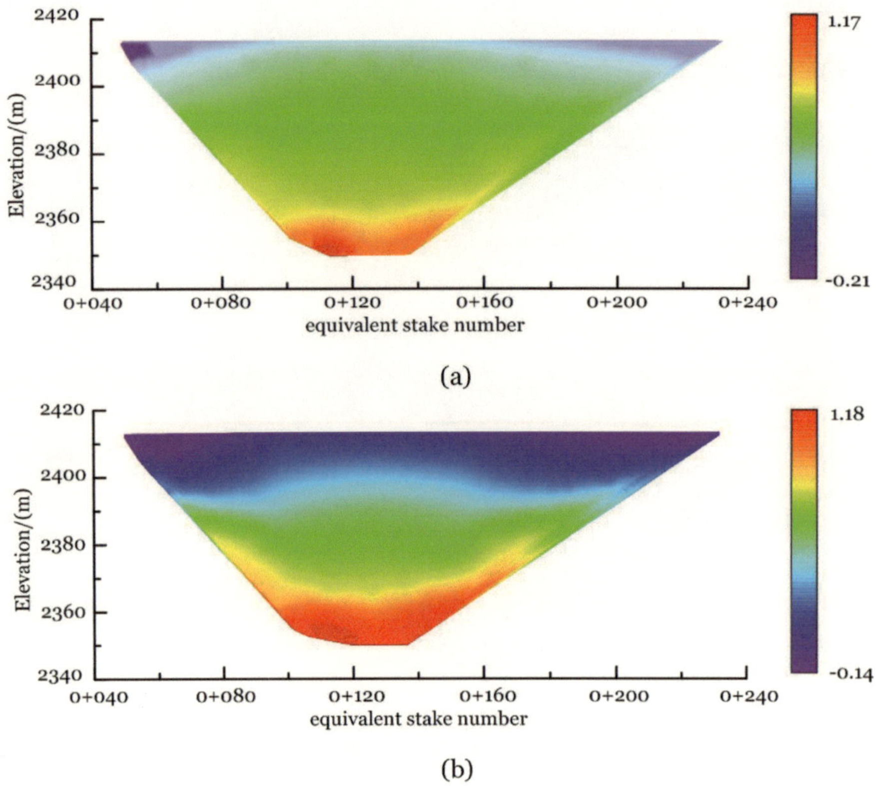

Figure 12.
Contour of the dam axial stresses on the downstream and upstream sides of the cutoff wall during the completion period (MPa). (a) Downstream side and (b) upstream side.

Considering the ultimate compressive strain of 700 με and ultimate tensile strain of 100 με, the allowable compressive strength and tensile strength of C25 concrete are 19.6 MPa and −2.8 MPa, respectively. From the above calculation results, the tensile and compressive stresses of the cutoff wall are all within the allowable range for C25 plain concrete (**Figure 13**).

5.3 Stress and deformation of the connecting plate and toe slab

Table 4 lists the characteristic values of the stress and deformation of the connecting plate and toe slab during the storage period.

Figure 14 shows the contour of the deformation of the connecting plate and toe slab during the storage period. For the axial displacement of the dam, the water displacement is represented by the compression from both sides of the riverbed. The axial displacement of the dam is generally small. The maximum displacements of the left and right banks after water storage are 0.71 cm and 0.89 cm, respectively, which occur in the 0 + 209 and 0 + 65 sections. The displacement of the river is characterized by a downward-directed deformation under the water load during the storage period, with a maximum value of 5.36 cm, which occurs in the 0 + 125 section of the riverbed. For the vertical displacement, the maximum value is 5.63 cm during the storage period, which also occurs at the 0 + 125 section of the riverbed. It can also be seen from **Figure 14** that due to the deformation joint between the connecting plate and the toe slab, the connection between the toe slab and the connecting plate is staggered, but the magnitude is small, and the setting of the toe slab length is appropriate.

DOI: http://dx.doi.org/10.5772/intechopen.92221

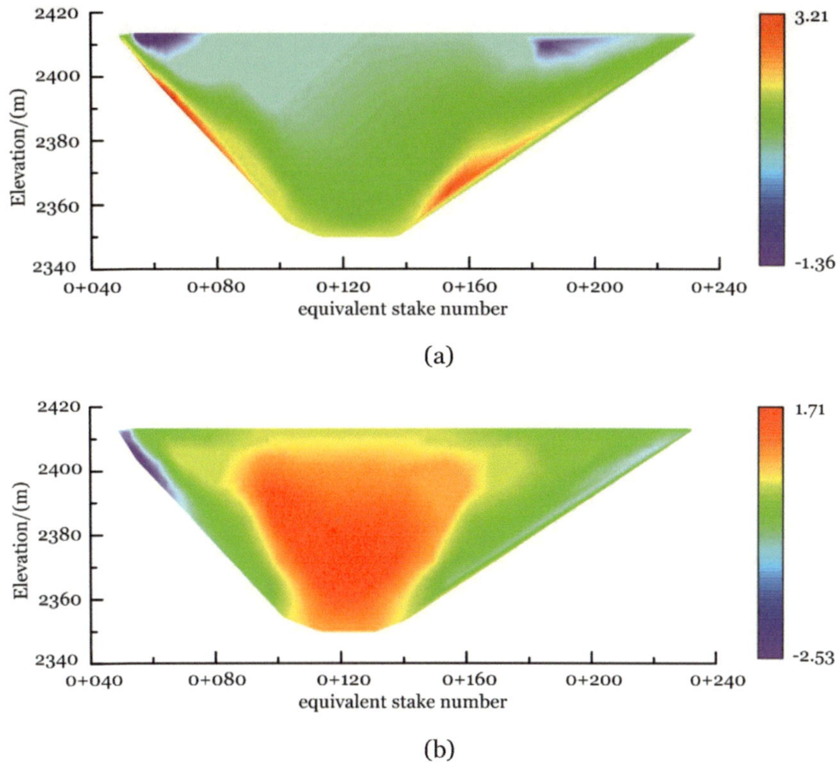

Figure 13.
Contour of the dam axial stresses on the downstream and upstream surface of the cutoff wall during the storage period (MPa). (a) Downstream side and (b) upstream side.

	Statistical items		Storage period
Connecting plate and toe slab	Dam axial displacement (cm)	Left side bank	−0.71
		Right side bank	0.89
	Displacement along the river (cm)	Upstream	/
		Downstream	5.36
	Settlement (cm)		5.63
Connecting plate	Dam axial stress (MPa)	Tensile stress	−1.81
		Compressive stress	0.56
	Major principal stress (MPa)	Compressive stress	0.86
	Minor principal stress (MPa)	Tensile stress	−1.82
		Compressive stress	0.32
Toe slab	Dam axial stress (MPa)	Tensile stress	−4.78
		Compressive stress	1.53
	Major principal stress (MPa)	Compressive stress	6.33
	Minor principal stress (MPa)	Tensile stress	−4.80
		Compressive stress	0.90

Table 4.
The characteristic values of the stress and deformation of the connecting plate and toe slab during the storage period.

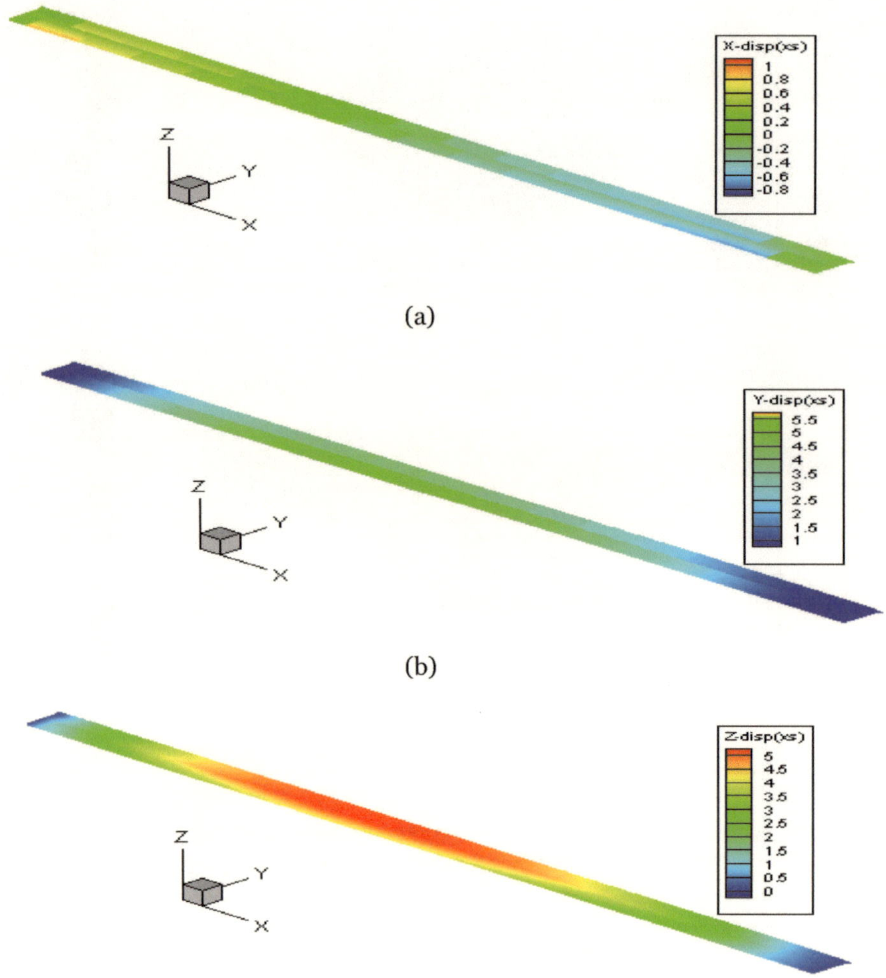

Figure 14.
Contour of the deformation of the connecting plate and toe slab during the storage period (cm). (a) Dam axial displacement, (b) displacement along the river and (c) settlement.

Figure 15.
Contour of the dam axial stresses of the connecting plate and toe slab during the storage period (MPa).

Figure 15 shows the contour of the dam axial stresses of the connecting plate and toe slab during the storage period. Under the action of water loading, the deformation of the connecting plate is constrained by the toe slab, and the deformation of the toe slab is constrained by the face slab, so the stress of the toe slab is greater than that of the connecting plate. The dam axial stress, corresponding to the deformation direction, is mainly manifested as tension at both ends and compression in the middle, and the downstream compressive stress is greater than the upstream compressive stress. After the storage period, the maximum tensile compressive stress is −4.78 MPa and 1.53 MPa, respectively, which occurs at the right end of the toe slab and in the 0 + 95 section of the riverbed.

Considering the ultimate compressive strain of 700 με and ultimate tensile strain of 100 με for C30 concrete, the allowable compressive strength and tensile strength are 27.3 MPa and −3.9 MPa, respectively. It can be seen from the above calculation results that the compressive stress and tensile stress of the connecting plate and toe slab are within the allowable range for C30 plain concrete, but the maximum tensile stress of the toe plate exceeds the allowable value of C30 plain concrete, and the exceeding area is mainly located in the local area at the junction of the toe slab and the bank slope, which could be resolved by adding reinforcement.

6. Conclusions

The Zhushou Reservoir was transformed from a clay core rock-debris dam to a concrete-faced rockfill dam, with the maximum dam height increasing from 63.4 m to 98.1 m. The three-dimensional finite element method was used to simulate the operation process of construction filling and the storage period, and the conclusions are discussed as follows:

1. The simulation results show that the maximum horizontal displacement occurs in the dam body of the old dam, and the maximum settlement occurs at the interface between the old and new dams. Due to the large modulus of cutoff wall and pile foundation, significant stress concentration has occurred in the dam.

2. During the storage period, the maximum axial tensile and compressive stresses of the cutoff wall are −2.53 MPa and 3.21 MPa, respectively, and the maximum major and minor principal stresses are 12.0 MPa and −1.74 MPa, respectively. The tensile and compressive stresses are all within the allowable range for C25 plain concrete, and the cutoff wall will not be damaged under static conditions.

3. During the storage period, the maximum axial tensile and compressive stresses of the toe slab (connecting plate) dam are −4.78 MPa and 1.53 MPa, respectively, and the maximum major and minor principal stresses are 6.33 MPa and −4.80 MPa, respectively. The compressive stress of toe slab and connecting plate and the tensile stress of connecting plate are all within the allowable range for C30 plain concrete, but the tensile stress of the local area at the junction of toe slab and bank slope has exceeded the allowable value for C30 plain concrete, so the reinforcement should be strengthened at this location.

Author details

Zhu Yumeng[1*], Guoying Li[1,2], Zhankuan Mi[1,2], Zhongzhi Fu[1,2]
and Kuangmin Wei[1,2]

1 Geotechnical Engineering Department, Nanjing Hydraulic Research Institute, Nanjing, China

2 Key Laboratory of Failure Mechanism and Safety Control Techniques of Earth-Rock Dam, Ministry of Water Resources, Nanjing, China

*Address all correspondence to: asbeel@163.com

References

[1] Hariri-Ardebili MA, Mirzabozorg H. Feasibility study of Dez arch dam heightening based on nonlinear numerical analysis of existing dam. Archives of Civil Engineering. 2013;**59**: 21-49. DOI: 10.2478/ace-2013-0002

[2] Weng Z. Study on the anti-seepage measures and slope stability during the heightening and strengthening of earth-fock fill dams. Journal of Zhejiang University of Water Resources and Electric Power. 2015;**27**:17-20, 26. DOI: 10.3969/j.issn.1008-536X.2015.02.004

[3] Lu Y, Xia S, Yue Y, Zhang J. Key technology for dam heightening of Songyue RCC dam. Chinese Journal of Geotechnical Engineering. 2008;**30**: 1614-1619. DOI: 10.3321/j.issn: 1000-4548.2008.11.007

[4] Zhang G, Zhu B, Zhipeng W. Temperature stress of gravity dam heightening. Journal of Hydraulic Engineering. 2003;**34**:11-15. DOI: 10.3321/j.issn:0559-9350.2003.05.002

[5] Schleiss AJ, Boes RM. Dams and Reservoirs Under Changing Challenges. London: CRC Press; 2011. p. 906. DOI: 10.1201/b11669

[6] Clerc B, Cesare GDE, Manso P. Heightening of very high gravity dams: The case study of the Grande Dixence. In: SCCER-SoE Annual Conference 2019-Hydropower and Geo-Energy in Switzerland: Challenges and Perspectives; 3–4 September 2019; Lausanne. Lausanne: SCCER-SoE; 2019. p. 134

[7] Morris SS, Garrett WS. The raising and strengthening of the Steenbras Dam. Civil Engineering = Siviele Ingenieurswese. 1956;**6**:137-149. DOI: 10.1680/iicep.1956.11454

[8] Daoud OMA, Sagady HS. Production and properties of high strength concrete for heightening concrete dam in Sudan. International Journal of GEOMATE: Geotechnique, Construction Materials and Environment. 2013;**4**:539-545. DOI: 10.21660/2013.8.24b

[9] Xiao H, Cui J, Xu Y. Research on the problems of bounding state between fresh and old concrete in the Dan Jiangkou Dam Heightening Project. South-to-North Water Transfers and Water Science & Technology. 2007;**5**: 8-11, 30. DOI: 10.13476/j.cnki.nsbdqk.2007.05.005

[10] Lu S, Zhang P. Dam heightening and danger eliminating design of Zhushou reservoir expansion and water diversion project. China Science and Technology Information. 2014;**21**:88-91. DOI: 10.3969/j.issn.1001-8972.2014.21.029

[11] Mi Z, Liu Y, Shi L, Zhang Z, Ying L. Feasibility of heightening concrete face rockfill dam on soft soil foundation. Journal of Water Resources and Architectural Engineering. 2017;**15**(1): 52-59. DOI: 10.3969/j.issn.1672-1144.2017.01.011

Chapter 3

Space-Time Finite Element Method for Seismic Analysis of Concrete Dam

Vikas Sharma, Akira Murakami and Kazunori Fujisawa

Abstract

Finite element method (FEM) is the most extended approach for analyzing the design of the dams against earthquake motion. In such simulations, time integration schemes are employed to obtain the response of the dam at time t_{n+1} from the known response at time t_n. To this end, it is desirable that such schemes are high-order accurate in time and remain unconditionally stable large time-step size can be employed to decrease the computation cost. Moreover, such schemes should attenuate the high-frequency components from the response of structure being studied. Keeping this in view, this chapter presents the theory of time-discontinuous space-time finite element method (ST/FEM) and its application to obtain the response of dam-reservoir system to seismic loading.

Keywords: space-time FEM, seismic response, concrete dam, time-integration, earthquake simulation

1. Introduction

During an event of earthquake stability of dams is of paramount importance as their failure can cause immense property and environmental damages. When dam-reservoir-foundation system is subjected to the dynamic loading it causes a coupled phenomenon; ground motion and deformations in the dam generate hydrodynamic pressure in the reservoir, which, in turn, can intensify the dynamic response of the dam. Moreover, spatial-temporal variation of stresses in the dam-body depends on the dynamic interactions between the dam, reservoir, and foundation. Therefore, it becomes necessary to use numerical techniques for the safety assessment of a given dam-design against a particular ground motion.

Dynamic finite element method is the most extended approach for computing the seismic response of the dam-reservoir system to the earthquake loading [1]. In this approach finite elements are used for discretization of space domain, and basis functions are locally supported on the spatial domain of these elements and remain independent of time. Furthermore, nodal values of primary unknowns depend only on time. Accordingly, this arrangement yields a system of ordinary differential equations (ODEs) in time which is then solved by employing time-marching schemes based on the finite difference method (FDM), such as Newmark-β method, HHT-α method, Houbolt method, and Wilson-θ method.

In dynamic finite element method (FEM), it is desirable to adopt large time-steps to decrease the computation time while solving a transient problem.

Therefore, it is imperative that the time-marching scheme remains unconditionally stable and higher order accurate [1]. In addition, it should filter out the high frequency components from the response of structure. To achieve these goals, Hughes and Hulbert presented space-time finite element method (ST/FEM) for solving the elastodynamics problem [2]. In this method, displacements (**u**) and velocities **v** are continuous in space-domain, however, discontinuous in time-domain. Li and Wiberg incorporated the time-discontinuity jump of displacements and velocities in the total energy norm to formulate an adaptive time-stepping ST/FEM [3]. ST/FEM, so far, has been successfully employed for solving linear and nonlinear structural dynamics problems [4, 5], moving-mass problems [6], and dynamical analysis of porous media [7], among other problems.

However, for elastodynamics problem, ST/FEM, yields a larger system of linear equations due to due to the time-discontinuous interpolation of displacement and velocity fields. Several efforts have been made in the past to overcome this issue; both explicit [8] and implicit [9] predictor-multi-corrector iteration schemes have been proposed to solve linear and nonlinear dynamics problems. Recently, to reduce the number of unknowns in ST/FEM, a different approach is taken in which only velocity is included in primary unknowns while displacement and stresses are computed from the velocity in a post-processing step [4, 10]. To this end, the objective of the present chapter is to introduce this method (henceforth, ST/FEM) in a pedagogical manner. The rest of the chapter is organized as follows. Sections 2 and 3 deal with the fundamentals of time-discontinuous Galerkin method. Section 4 describes the dam-reservoir-soil interaction problem, and Section 5 discusses the application of ST/FEM for this problem. Lastly, Section 6 demonstrates the numerical performance of proposed method and in the last section concluding remarks are included.

2. Time-discontinuous Galerkin method (tDGM) for second order ODE

Consider a mass-spring-dashpot system as depicted in **Figure 1**. The governing equation of motion is described by the following second order initial value problem in time.

$$
\begin{aligned}
\frac{d^2u}{dt^2} + 2\zeta\omega_n\frac{du}{dt} + \omega_n^2 u &= f(t) \quad \forall t \in [0, T] \\
u(0) &= u_0 \\
\frac{du(0)}{dt} &= v_0
\end{aligned}
\tag{1}
$$

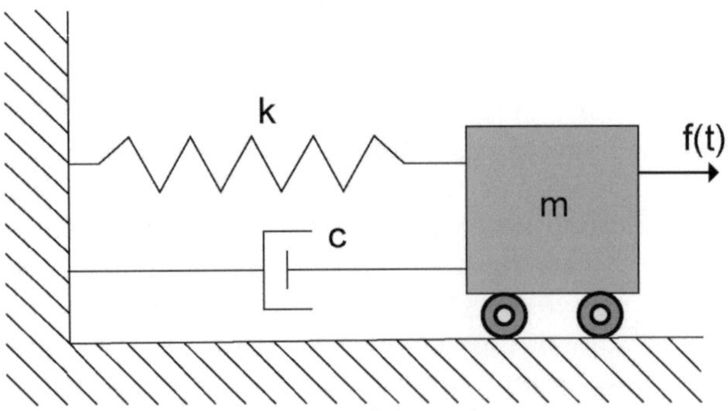

Figure 1.
Schematic diagram of the mass-spring-dashpot system.

where $u := u(t)$ is the unknown displacement, $f(t)$ is the external force acting on the system. Further, u_0 and v_0 are the prescribed initial values of the displacement and velocity, respectively. Damping ratio ζ and the natural frequency of vibration ω_n of the system are related to the mass m, stiffness of the spring k, and damping coefficient c by:

$$\omega_n = \sqrt{k/m}, \qquad \zeta = \frac{c}{2m\omega_n} = \frac{c}{2\sqrt{mk}} \tag{2}$$

In what follows, this second order ODE will be utilized to discuss the fundamental concepts behind time-discontinuous Galerkin methods (henceforth, tDGM).

2.1 Two-field tDGM

In two-field tDGM (henceforth, uv-tDGM), both displacement (u) and velocity (v) are treated as independent primary variables and interpolated by using the piecewise polynomials. Both u and v are discontinuous at end-points (i.e., t_n and t_{n+1}) of time-slab $I_n = (t_n, t_{n+1})$. However, u and v remain continuous inside I_n, and approximated by piecewise polynomials (refer, **Figure 2**). Therefore, discontinuity occurs at discrete times belonging to a set $\{t_0, t_1, \cdots, t_N\}$. The jump discontinuity in time for u is denotes by

$$[\![u]\!]_n = u_n^+ - u_n^- \tag{3}$$

where

$$u_n^+ = \lim_{\varepsilon \to 0} u(t+\varepsilon), \qquad u_n^- = \lim_{\varepsilon \to 0} u(t-\varepsilon) \tag{4}$$

are the discontinuous values of u at time $t = t_n$. By recasting Eq. (1) into a system of two first-order ODEs one can obtain,

$$\frac{dv}{dt} + 2\zeta\omega_n v + \omega_n^2 u = f(t) \quad \forall t \in [0, T] \tag{5}$$

$$\frac{du}{dt} - v = 0 \quad \forall t \in [0, T] \tag{6}$$

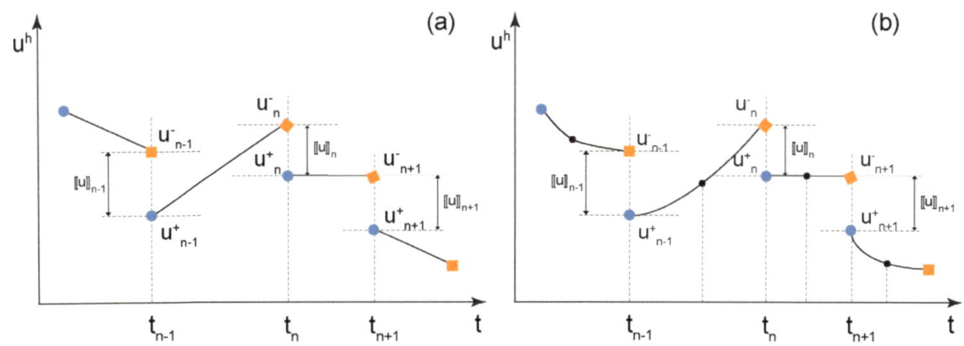

Figure 2.
Schematic diagram of time discontinuous approximation: (a) piecewise linear interpolation, and (b) piecewise quadratic interpolation.

$$u(0) = u_0, \quad v(0) = v_0 \tag{7}$$

The weak-form of the uv-tDGM can be stated as: find $u^h \in \mathfrak{I}_l^h$ and $v^h \in \mathfrak{I}_l^h$, such that for all $\delta u^h \in \mathfrak{I}_l^h$ and $\delta v^h \in \mathfrak{I}_l^h$, and for all $n = 0, \cdots, N-1$ Eq. (8) holds.

$$\begin{aligned}&\int_{I_n} \delta v^h \left(\frac{dv^h}{dt} + 2\zeta\omega_n v^h + \omega_n^2 u^h - f(t)\right) dt + \delta v^h(t_n)\left[\left[v^h\right]\right]_n \\ &+ \int_{I_n} \delta u^h \left(\frac{du^h}{dt} - v^h\right) dt + \delta u^h(t_n)\left[\left[u^h\right]\right]_n = 0\end{aligned} \tag{8}$$

where $\mathfrak{I}_l^h$ denotes the collection of polynomial with order less than or equal to l. It is worth noting that the presence $\left[\left[u^h\right]\right]_n$ and $\left[\left[v^h\right]\right]_n$ correspond to the weakly enforced initial condition for the u and v, respectively. Further, since the selection of δu^h and δv^h is arbitrary one can depict Eq. (8) as,

$$\int_{I_n} \delta v^h \left(\frac{dv^h}{dt} + 2\zeta\omega_n v^h + \omega_n^2 u^h - f(t)\right) dt + \delta v^h(t_n)\left[\left[v^h\right]\right]_n = 0 \tag{9}$$

$$\int_{I_n} \delta u^h \left(\frac{du^h}{dt} - v^h\right) dt + \delta u^h(t_n)\left[\left[u^h\right]\right]_n = 0 \tag{10}$$

Eq. (10) denotes that, in uv-tDGM, displacement-velocity compatibility relationship is satisfied in weak form.

2.2 Single field tDGM

To decrease the number of unknowns in comparison to those involved in uv-tDGM, displacement-velocity compatibility condition (cf. Eq. 6) can be explicitly satisfied and velocity can be selected as primary unknown. Henceforth, this strategy will be termed as v-tDGM. In v-tDGM, v is continuous in I_n, but discontinuity occurs at the end-points t_n, t_{n+1}. Further, u is computed in a post-processing step by integration of v, therefore, u remains continuous in time $[0, T]$.

The weak form of the v-tDGM reads: Find $v^h \in \mathfrak{I}_l^h$ such that for all $\delta v^h \in \mathfrak{I}_l^h$, and for all $n = 0, \cdots, N-1$ Eq. (11) holds.

$$\int_{I_n} \delta v^h \left(\frac{dv^h}{dt} + 2\zeta\omega_n v^h + \omega_n^2 u^h - f(t)\right) dt + \delta v^h(t_n)\left[\left[v^h\right]\right]_n = 0 \tag{11}$$

Note that Eqs. (9) and (11) are identical, however, in former, u^h is an independent variable and, in later, it is a dependent variable which will be computed by using following expression.

$$u^h(t) = u(t_n) + \int_{t_n}^{t} v^h(\tau) d\tau \tag{12}$$

Let us now focus on the discretization of weak-form (cf. Eq. (11)) by using the locally defined piecewise linear test and trial functions,

$$v^h = T_1 v_n^+ + T_2 v_{n+1}^- \qquad \delta v^h = T_1 \delta v_n^+ + T_2 \delta v_{n+1}^- \tag{13}$$

$$u^h(t) = u_n + v_n^+ \frac{\Delta t_n}{2}\left(1 - T_1^2\right) + v_{n+1}^- \frac{\Delta t_n}{2} T_2^2 \tag{14}$$

where

$$T_1(\theta) = \frac{1-\theta}{2} \qquad T_2(\theta) = \frac{1+\theta}{2}, \qquad \theta \in [-1, 1]. \tag{15}$$

Accordingly, Eq. 11 transforms into following matrix-vector form.

$$\begin{gathered}\frac{1}{2}\begin{bmatrix} 1 & 1 \\ -1 & 1 \end{bmatrix}\begin{Bmatrix} v_n^- \\ v_{n+1}^+ \end{Bmatrix} + \frac{2\zeta\omega_n \Delta t_n}{6}\begin{bmatrix} 2 & 1 \\ 1 & 2 \end{bmatrix}\begin{Bmatrix} v_n^- \\ v_{n+1}^+ \end{Bmatrix} \\ + \frac{\omega_n^2 \Delta_n^2}{24}\begin{bmatrix} 3 & 1 \\ 5 & 3 \end{bmatrix}\begin{Bmatrix} v_n^- \\ v_{n+1}^+ \end{Bmatrix} = \begin{Bmatrix} J_{ext}^1 \\ J_{ext}^2 \end{Bmatrix} - \frac{\omega_n^2 \Delta t_n u_n}{2}\begin{Bmatrix} 1 \\ 1 \end{Bmatrix} + \begin{Bmatrix} v_n^- \\ 0 \end{Bmatrix}\end{gathered} \tag{16}$$

where J_{ext}^1 and J_{ext}^2 are given by

$$J_{ext}^1 = \int_{I_n} T_1 f(t)dt \qquad J_{ext}^2 = \int_{I_n} T_2 f(t)dt$$

3. Numerical analysis of tDGM

In this section, numerical analysis of the tDGM schemes, (viz. uv-tDGM and v-tDGM) for the second order ODE will be performed. To assess the stability characteristics and temporal accuracy of these schemes, classical finite difference techniques will be used ([11], Chapter 9). In this context, it is sufficient to consider the following homogeneous and undamped form of Eq. (1):

$$\frac{d^2u}{dt^2} + \omega_n^2 u = 0, \qquad u(0) = u(0), \qquad \frac{du(0)}{dt} = v_0, \qquad t \in [0, T] \tag{17}$$

3.1 Energy decay in v-tDGM

In this section it will be shown that v-tDGM is a true energy-decaying scheme. Consider Eq. (17) which represents the governing equation of a spring-mass system. The total energy (sum of kinetic and potential energy) of the system remains constant because damping and external forces are absent in the system.

$$T_E(u, v) := \frac{1}{2}v^2 + \frac{1}{2}\omega_n^2 u^2 = \text{constant} \tag{18}$$

Consider the time domain $[0, T]$ and corresponding N time-slabs; $I_n := (t_n, t_{n+1})$ for $n = 0, 1 \cdots, N-1$. Let the u and v at time $t_0 = 0$ be given by $u_0^+ = u_0^- = u_0$, and $v_0^+ = v_0^- = v_0$, respectively. Furthermore, the u and v at time $t_N = T$ are denoted by u_N^- and u_N^-, respectively.

Accordingly, it can be shown that

$$T_E\left(u_N, v_N^-\right) = T_E(u_0, v_0) - \frac{1}{2}\sum_{n=0}^{N-1} \left[\left[v^h\right]\right]_n^2$$

or

$$T_E\left(u_N, v_N^-\right) \leq T_E(u_0, v_0)$$

This shows that v-tDGM is an energy decaying time integration algorithm, in which the total energy during any time step, $T_E\left(u_N^-, v_N^-\right)$, is always bounded from above by the total energy at the first time-step (i.e., $T_E(u_0, v_0)$).

To assess the energy dissipation characteristics of v-tDGM, Eq. (17) is solved with $\omega_n = 2\pi$, $u_0 = 0$, and $v_0 = 1.0 m/s$. The undamped time period T_0 of the sinusoidal motion is 1.0 second, and the total time duration of simulation is $T = 50$ seconds. **Figure 3a** depicts the time history graphs of the normalized total energy (i.e., $T_E(u, v)/T_E(u_0, v_0)$) computed by using v-tDGM with different time step sizes. Further, to visualize the effect of energy-dissipation displacement-velocity phase diagram is plotted in **Figure 3b**. For present problem, phase-diagram should be an ellipse. The presence of energy dissipation in the numerical algorithm, however, decreases the total energy which results in shortening of the radius of ellipse. From these plots it is evident that the dissipation of energy decreases as the time-step size decreases which also indicates that the jump discontinuity in time decreases with time-step size.

3.2 Stability characteristics of v-tDGM

In this section, to study the stability characteristics of v-tDGM, Eq. (17) is considered. The matrix-vector form corresponding to this problem is given by

$$\frac{1}{2}\begin{bmatrix} 1 & 1 \\ -1 & 1 \end{bmatrix}\begin{Bmatrix} v_n^- \Delta t_n \\ v_{n+1}^+ \Delta t_n \end{Bmatrix} + \frac{\Omega^2}{24}\begin{bmatrix} 3 & 1 \\ 5 & 3 \end{bmatrix}\begin{Bmatrix} v_n^- \Delta t_n \\ v_{n+1}^+ \Delta t_n \end{Bmatrix} = \begin{Bmatrix} v_n^- \Delta t_n \\ 0 \end{Bmatrix} - \begin{Bmatrix} \dfrac{\Omega^2 u_n}{2} \\ \dfrac{\Omega^2 u_n}{2} \end{Bmatrix}, \tag{19}$$

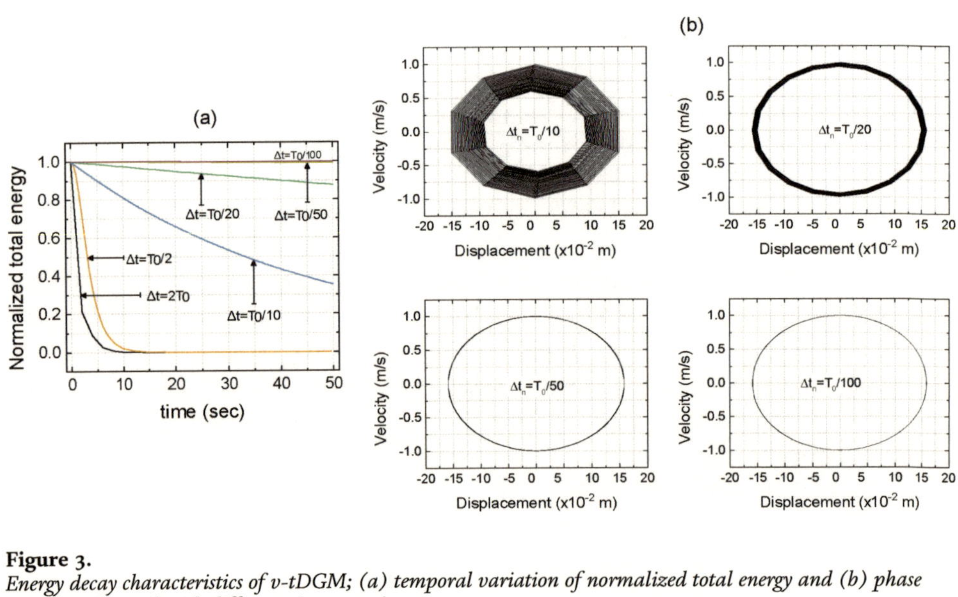

Figure 3.
Energy decay characteristics of v-tDGM; (a) temporal variation of normalized total energy and (b) phase diagram obtained with different time-step sizes.

where $\Omega = \omega_n \Delta t_n$. Subsequently, eliminating v_n^+ in Eq. (19),

$$\left\{ \begin{array}{c} u_{n+1} \\ v_{n+1}^- \Delta t_n \end{array} \right\} = \mathbf{A}(\Omega) \left\{ \begin{array}{c} u_n \\ v_n^- \Delta t_n \end{array} \right\}, \tag{20}$$

where $\mathbf{A}$ is the amplification matrix given by,

$$\mathbf{A}(\Omega) = \begin{bmatrix} \dfrac{\Omega^4 - 30\Omega^2 + 72}{\Omega^4 + 6\Omega^2 + 72} & \dfrac{-6\Omega^2 + 72}{\Omega^4 + 6\Omega^2 + 72} \\ \\ \dfrac{6\Omega^4 - 72\Omega^2}{\Omega^4 + 6\Omega^2 + 72} & \dfrac{-30\Omega^2 + 72}{\Omega^4 + 6\Omega^2 + 72} \end{bmatrix}. \tag{21}$$

To investigate the stability of v-tDGM one should look into the eigenvalues of $\mathbf{A}$ (here, denoted by λ_1 and λ_2). Let the modulus of λ be denoted by $|\lambda| = \sqrt{\lambda\lambda^*}$ with λ^* denoting the complex conjugate of λ. Accordingly, the spectral radius of $\mathbf{A}$ can be described by $\rho(\mathbf{A}) = \max_{i=1,2} |\lambda_i(\mathbf{A})|$. It can be easily shown that v-tDGM satisfies all criteria for the spectral stability [11, Chapter 9]: (a) $\rho \leq 1$, (b) eigenvalues of $\mathbf{A}$ of multiplicity greater than one are strictly less than one in modulus. It proves that v-tDGM is an unconditionally stable time-marching scheme (for more details, readers are referred to [4]).

3.3 High-frequency response of TDG/FEM

Figure 4 plots the frequency responses of $\rho(\mathbf{A})$ for v-tDGM. It is evident that $\rho \leq 1$ which proves that present algorithm is unconditionally stable. The v-TDG/FEM, however, cannot attenuate spurious high-frequency contents since $\rho_\infty = 1$ (see **Figure 4**). However, v-tDGM provides negligible attenuation in the small frequency regime as ρ is close to one in this regime.

3.4 Accuracy of v-tDGM

In [4], it is shown that u in Eq. (20) satisfies the following finite difference stencil.

$$u_{n+1} - 2a_1 u_n + a_2 u_{n-1} = 0, \tag{22}$$

where $a_1 = Trace\mathbf{A}/2$ and $a_2 = det\mathbf{A}$. Let us now denote the exact solutions by $u(t)$ and $v(t)$. Then the local truncation error $\tau(t)$ corresponding to Eq. (22) at any time t becomes

$$u(t + \Delta t) - 2a_1 u(t) + a_2 u(t - \Delta t) = \Delta t^2 \tau(t) \tag{23}$$

Subsequently, by expanding $u(t + \Delta t)$ and $u(t - \Delta t)$ about t by using Taylor series, and by using Eq. (17), it can be proved that v-tDGM is consistent and third order accurate, i.e., $|\tau(t)| \leqslant \frac{1}{72} \Delta t^3$ [4]. Accordingly, one can use the Lax equivalence theorem to prove the convergence of the algorithms.

A direct consequence of the convergence is that the solution of Eq. (17) can be given by following expression [1]:

$$u_n = \exp\left(-\frac{\overline{\zeta}\,\overline{\Omega} t_n}{\Delta t}\right)\left[k_1 \cos\left(\frac{\overline{\Omega} t_n}{\Delta t}\right) + k_2 \sin\left(\frac{\overline{\Omega} t_n}{\Delta t}\right)\right] \tag{24}$$

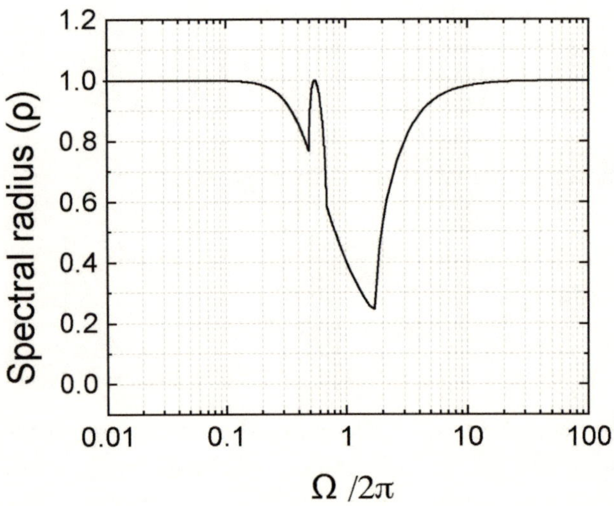

Figure 4.
Frequency response of spectral radius ρ for v-tDGM.

with

$$\overline{\Omega} = \arctan\left(\frac{\sqrt{a_2 - a_1^2}}{a_1}\right) \qquad \overline{\zeta} = -\frac{1}{2\overline{\Omega}}\ln(a_2) \tag{25}$$

where $\overline{\zeta}$ denotes the algorithmic damping ratio, $\overline{\Omega}$ is the frequency of the discrete solutions, and the coefficients k_1 and k_2 are determined by the displacement and velocity initial conditions.

Further, to investigate the accuracy of v-tDGM, algorithmic damping ratio, which is a measure of amplitude decay, and relative frequency error $(\Omega - \overline{\Omega})/\overline{\Omega}$, which is a measure of relative change in time period, are plotted in **Figure 5**. From **Figure 5a** it can be observed that ζ is comparable with the HHT-α scheme, however, it is significantly smaller than the uv-tDGM. It is evident that the Houbolt and Wilson-θ methods are too dissipative in the low-frequency range, therefore, these algorithms are not suitable for the long-duration numerical simulations. Furthermore, v-tDGM has smallest frequency error which can be attributed to its third

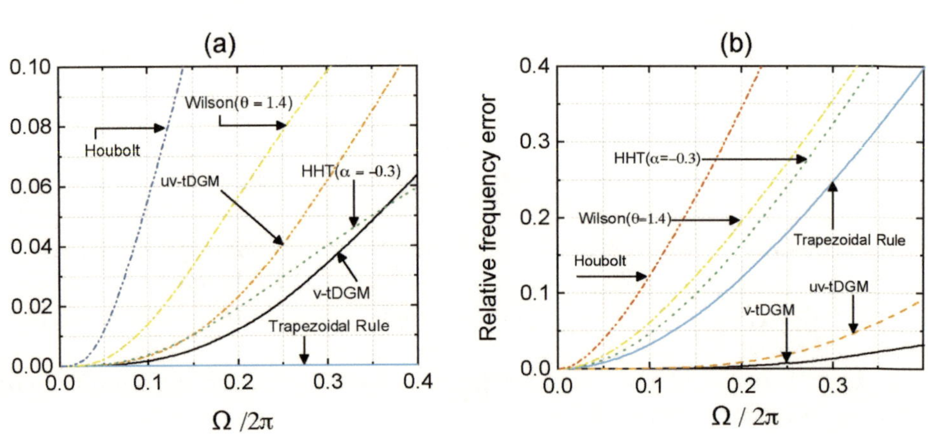

Figure 5.
Accuracy of v-tDGM: (a) algorithmic damping ratio, and (b) relative frequency error in low frequency regime (after [4]).

order accuracy (refer, **Figure 5b**). It can be stated that these characteristics of v-tDGM, such as very low numerical dispersion and dissipation, third-order accuracy, and unconditional stability, make this scheme suitable for long-time simulations. However, at present, the only possible drawback to this method is its incapability to attenuate the spurious high-frequency components.

4. Statement of problem

A dam-reservoir-soil (DRS) system which is subjected to the spatially uniform horizontal ($a_1^g(t)$) and vertical ($a_2^g(t)$) component of ground motion is depicted in **Figure 6**. Reservoir domain contains linear, inviscid, irrotational, and compressible fluid and solid domain (dam and underlying soil) is treated as isotropic, homogeneous, linear elastic material. Computation domain of soil (Ω^s) and fluid (Ω^f) are obtained by prescribing the viscous boundary conditions at the artificial boundaries [10]. Let Γ_f^f and Γ_∞^f be the free surface and upstream artificial boundary of fluid domain. Γ_{fs}^f, Γ_{fd}^f, Γ_{fd}^s and Γ_{fs}^s denote the fluid-soil, fluid-dam, dam-fluid and soil-fluid interfaces, respectively. Further, the outward unit normal vectors to the fluid and solid boundary are given by $\mathbf{n}^s$ and $\mathbf{n}^f$, respectively.

Further, hydrodynamic pressure distribution in the reservoir is modeled by the pressure wave equation,

$$\frac{1}{c^2}\frac{\partial^2 p}{\partial t^2} - \nabla^2 p = 0 \quad \text{in} \quad \Omega^f \quad \forall t \in (0, T), \tag{26}$$

with following initial and boundary conditions.

$$p(\mathbf{x}, 0) = 0; \frac{\partial p(\mathbf{x}, 0)}{\partial t} = 0 \quad \text{in} \quad \Omega^f \quad at \quad t = 0 \tag{27}$$

$$p(\mathbf{x}, t) = 0 \quad \text{on} \quad \Gamma_f^f \quad \forall t \in (0, T) \tag{28}$$

$$\nabla p \cdot \mathbf{n}^f = -\rho^f \frac{\partial \mathbf{v}}{\partial t} \cdot \mathbf{n}^f \quad \text{on} \quad \Gamma_{fd}^f \cup \Gamma_{fs}^f \quad \forall t \in (0, T) \tag{29}$$

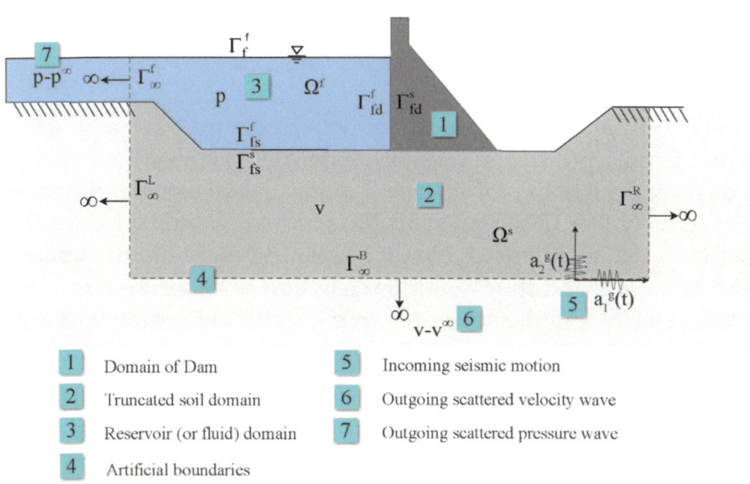

Figure 6.
Schematic diagram of dam-reservoir-soil (DRS) system subjected to seismic ground motion.

$$\nabla p \cdot \mathbf{n}^f = -\frac{1}{c}\frac{\partial p}{\partial t} + \frac{1}{c}\frac{\partial p_\infty}{\partial t} \quad \text{on} \quad \Gamma_\infty^f \quad \forall t \in (0, T) \tag{30}$$

In Eq. (26), ∇^2 denotes the Laplace's operator, $p(\mathbf{x}, t)$ denotes the hydrodynamic pressure in the water (in excess of hydrostatic pressure) and c denotes the speed of sound in water. Eq. (29) denotes the time dependent boundary condition at fluid-dam and fluid-soil interface, respectively, where ρ^f is the mass density of fluid, and $\mathbf{v}$ is the velocity of solid domain (i.e., dam or soil). Eq. (30) is due to the viscous boundary condition at the upstream truncated boundary of reservoir. The first term in this equation corresponds to an array of dashpots placed normal to the truncated boundary Γ_∞^f, and the second term is due to the free-field response of reservoir.

Let us now consider the initial-boundary value problem of the solid domain which is described by,

$$\rho^s \frac{\partial^2 \mathbf{u}}{\partial t^2} - \nabla \cdot \boldsymbol{s} - \rho^s \mathbf{b} = 0 \quad \text{in} \quad \Omega^s \quad \forall t \in (0, T), \tag{31}$$

$$\mathbf{u}(\mathbf{x}, t) = \mathbf{g}(\mathbf{x}, t) \quad \text{on} \quad \Gamma^g \quad \forall t \in \times (0, T), \tag{32}$$

$$\boldsymbol{\sigma} \cdot \mathbf{n}^s = \mathbf{h} \quad \text{on} \quad \Gamma_\infty^+ \cup \Gamma_{fd}^s \cup \Gamma_{fs}^s \quad \forall t \in (0, T), \tag{33}$$

$$\mathbf{u}(\mathbf{x}, 0) = \mathbf{u}^0(\mathbf{x}), \quad \frac{\partial \mathbf{u}}{\partial t}(\mathbf{x}, 0) = \mathbf{v}^0(\mathbf{x}) \quad \text{in} \quad \Omega^s \quad \text{at} \quad t = 0. \tag{34}$$

Furthermore, following time dependent boundary conditions will be considered in Eq. (33):

$$\boldsymbol{\sigma} \cdot \mathbf{n}^s = -\mathbf{c}^v(\mathbf{v} - \mathbf{v}^\infty) + \boldsymbol{\sigma}^\infty \cdot \mathbf{n}^s \quad \text{on} \quad \Gamma_\infty^L \cup \Gamma_\infty^R \tag{35}$$

$$\boldsymbol{\sigma} \cdot \mathbf{n}^s = -\mathbf{c}^h \cdot \mathbf{v} + 2\mathbf{c}^h \cdot \mathbf{v}^{in} \quad \text{on} \quad \Gamma_\infty^B \tag{36}$$

$$\boldsymbol{\sigma} \cdot \mathbf{n}^s = -\{p_h(\mathbf{x}) + p(\mathbf{x}, t)\}\mathbf{n}^s \quad \text{on} \quad \Gamma_{fd}^s \cup \Gamma_{fs}^s \tag{37}$$

In addition, solid domain is considered to be an isotropic, homogeneous, linear elastic material with

$$\sigma_{ij} = \lambda \varepsilon_{kk}\delta_{ij} + 2\mu\varepsilon_{ij}, \tag{38}$$

$$\varepsilon_{ij} = \frac{1}{2}\left(\frac{\partial u_i}{\partial x_j} + \frac{\partial u_j}{\partial x_i}\right). \tag{39}$$

In Eqs. (31)–(34), ρ^s, $\mathbf{u}$, σ, $\mathbf{b}$, $\mathbf{g}$, $\mathbf{h}$, $\mathbf{u}^0$ and $\mathbf{v}^0$, denote mass density, displacement, Cauchy's stress tensor (positive in tension), externally applied body force density, prescribed displacement, external surface traction, initial value of displacement and velocity, respectively. Eq. (37) represents time varying boundary condition due to the hydrostatic, $p_h(\mathbf{x})$, and hydrodynamic, $p(\mathbf{x}, t)$, pressure of impounded water acting on the dam-fluid and fluid-soil interface. Furthermore, In Eqs. (35)–(37), $\mathbf{v}^\infty$ and σ^∞ are the velocity and the stress due to the free-field response of unbounded soil domain. In addition, Γ_∞^L, Γ_∞^R, and Γ_∞^B represent left, right and bottom truncated boundaries of soil domain, respectively. In Eqs. (35) and (36), first term corresponds to the Lysmer and Kuhlemeyer viscous boundary condition. Physically, it represents a series of dashpots placed at the truncated boundaries of soil domain in parallel and normal directions (see **Figure 6b**). Further, these terms facilitate the absorption of outgoing scattered wave motion and attempt to model the radiation damping due to the semi-infinite soil domain. Where $\mathbf{c}^v$ and $\mathbf{c}^h$ are the damping coefficients matrices for the dashpots placed at vertical and horizontal truncated boundaries:

$$\mathbf{c}^v = \begin{bmatrix} \rho^s c_L & 0 \\ 0 & \rho^s c_T \end{bmatrix}, \qquad \mathbf{c}^h = \begin{bmatrix} \rho^s c_T & 0 \\ 0 & \rho^s c_L \end{bmatrix}, \tag{40}$$

in which, c_L and c_T are the speed of longitudinal wave (P-wave) and transverse wave (S-wave) in the unbounded soil domain, respectively. Lastly, in Eqs. (38) and (39), λ and μ are the Lame parameters, and δ_{ij} is the Kronecker delta function.

5. Space-time finite element method

Recently, ST/FEM is employed to solve dam-reservoir-soil interaction problem [10] in which the resultant matrix-vector form is given by

$$\left[\mathbf{K}_{11}^f\right]\{\mathbf{Q}_1\} + \left[\mathbf{K}_{12}^f\right]\{\mathbf{Q}_2\} - \frac{1}{2}\left[\mathbf{H}^f\right]\{\mathbf{V}_1\} + \frac{1}{2}\left[\mathbf{H}^f\right]\{\mathbf{V}_2\} = \left\{\mathbf{J}_1^f\right\} \tag{41}$$

$$\left[\mathbf{K}_{21}^f\right]\{\mathbf{Q}_1\} + \left[\mathbf{K}_{22}^f\right]\{\mathbf{Q}_2\} - \frac{1}{2}\left[\mathbf{H}^f\right]\{\mathbf{V}_1\} + \frac{1}{2}\left[\mathbf{H}^f\right]\{\mathbf{V}_2\} = \left\{\mathbf{J}_2^f\right\} \tag{42}$$

$$\left[\mathbf{K}_{11}^s\right]\{\mathbf{V}_1\} + \left[\mathbf{K}_{12}^s\right]\{\mathbf{V}_2\} + \frac{3\Delta t_n^2}{24}\left[\mathbf{H}^s\right]\{\mathbf{Q}_1\} + \frac{\Delta t_n^2}{24}\left[\mathbf{H}^s\right]\{\mathbf{Q}_2\} = \left\{\mathbf{J}_1^s\right\} \tag{43}$$

$$\left[\mathbf{K}_{21}^s\right]\{\mathbf{V}_1\} + \left[\mathbf{K}_{22}^s\right]\{\mathbf{V}_2\} + \frac{5\Delta t_n^2}{24}\left[\mathbf{H}^s\right]\{\mathbf{Q}_1\} + \frac{3\Delta t_n^2}{24}\left[\mathbf{H}^s\right]\{\mathbf{Q}_2\} = \left\{\mathbf{J}_2^s\right\} \tag{44}$$

where $\{\mathbf{Q}_1\}$ and $\{\mathbf{Q}_2\}$ represent the spatial nodal values of auxiliary variable $q = \partial p/\partial t$ at time t_n^+ and t_{n+1}^-, respectively. Similarly, $\{\mathbf{V}_1\}$ and $\{\mathbf{V}_2\}$ are the spatial nodal values of velocity field at time t_n^+ and t_{n+1}^-, respectively (for more details see [10]).

Further, In Eqs. (41) and (42),

$$\left[\mathbf{K}_{11}^f\right] = \frac{1}{2}\left[\mathbf{M}^f\right] + \frac{3\Delta t_n^2}{24}\left[\mathbf{K}^f\right] + \frac{\Delta t_n}{3}\left[\mathbf{C}_\infty^f\right], \tag{45}$$

$$\left[\mathbf{K}_{12}^f\right] = \frac{1}{2}\left[\mathbf{M}^f\right] + \frac{\Delta t_n^2}{24}\left[\mathbf{K}^f\right] + \frac{\Delta t_n}{6}\left[\mathbf{C}_\infty^f\right], \tag{46}$$

$$\left[\mathbf{K}_{21}^f\right] = -\frac{1}{2}\left[\mathbf{M}^f\right] + \frac{5\Delta t_n^2}{24}\left[\mathbf{K}^f\right] + \frac{\Delta t_n}{6}\left[\mathbf{C}_\infty^f\right], \tag{47}$$

$$\left[\mathbf{K}_{22}^f\right] = \frac{1}{2}\left[\mathbf{M}^f\right] + \frac{3\Delta t_n^2}{24}\left[\mathbf{K}^f\right] + \frac{\Delta t_n}{3}\left[\mathbf{C}_\infty^f\right], \tag{48}$$

where

$$\left[\mathbf{M}^f\right] = \int_{\Omega^f} \mathbf{N}^T\mathbf{N}\frac{1}{c^2}\,d\Omega \tag{49}$$

$$\left[\mathbf{K}^f\right] = \int_{\Omega^f} \nabla\mathbf{N}\cdot\nabla\mathbf{N}^T\,d\Omega \tag{50}$$

$$\left[\mathbf{C}_\infty^f\right] = \int_{\Gamma_\infty^f} \mathbf{N}^T\mathbf{N}\frac{1}{c}\,ds \tag{51}$$

denote mass matrix, diffusion matrix, and viscous boundary at the upstream truncated boundary, respectively. The fluid-solid coupling matrix $\left[\mathbf{H}^f\right]$ is given by,

$$[\mathbf{H}^f] = \int_{\Gamma_{fs}^f \cup \Gamma_{fd}^f} \mathbf{N}^T \mathbf{N} \rho^f \mathbf{n}^f \, ds \tag{52}$$

and right hand side spatial nodal vectors in Eqs. (41) and (42) becomes,

$$\left\{\mathbf{J}_1^f\right\} = [\mathbf{M}^f]\{\mathbf{Q}_0\} - \frac{\Delta t_n}{2}[\mathbf{K}^f]\{\mathbf{P}_0\} + \frac{\Delta t_n}{3}[\mathbf{C}_\infty^f]\{\mathbf{Q}_1^\infty\} + \frac{\Delta t_n}{6}[\mathbf{C}_\infty^f]\{\mathbf{Q}_2^\infty\} \tag{53}$$

$$\left\{\mathbf{J}_2^f\right\} = -\frac{\Delta t_n}{2}[\mathbf{K}^f]\{\mathbf{P}_0\} + \frac{\Delta t_n}{6}[\mathbf{C}_\infty^f]\{\mathbf{Q}_1^\infty\} + \frac{\Delta t_n}{3}[\mathbf{C}_\infty^f]\{\mathbf{Q}_2^\infty\} \tag{54}$$

where $\{\mathbf{Q}_0\}$ and $\{\mathbf{P}_0\}$ correspond to the nodal values of q and p at time t_n, and $\{\mathbf{Q}_1^\infty\}$ and $\{\mathbf{Q}_2^\infty\}$ are the free field hydrodynamic response of reservoir at time t_n and t_{n+1}, respectively.

In Eqs. (43) and (44),

$$[\mathbf{K}_{11}^s] = \frac{1}{2}[\mathbf{M}^s] + \frac{3\Delta t_n^2}{24}[\mathbf{K}^s] + \frac{\alpha \Delta t_n}{3}[\mathbf{M}^s] + \frac{\beta \Delta t_n}{3}[\mathbf{K}^s] + \frac{\Delta t_n}{3}[\mathbf{C}_\infty^s], \tag{55}$$

$$[\mathbf{K}_{12}^s] = \frac{1}{2}[\mathbf{M}^s] + \frac{\Delta t_n^2}{24}[\mathbf{K}^s] + \frac{\alpha \Delta t_n}{6}[\mathbf{M}^s] + \frac{\beta \Delta t_n}{6}[\mathbf{K}^s] + \frac{\Delta t_n}{6}[\mathbf{C}_\infty^s], \tag{56}$$

$$[\mathbf{K}_{21}^s] = -\frac{1}{2}[\mathbf{M}^s] + \frac{5\Delta t_n^2}{24}[\mathbf{K}^s] + \frac{\alpha \Delta t_n}{6}[\mathbf{M}^s] + \frac{\beta \Delta t_n}{6}[\mathbf{K}^s] + \frac{\Delta t_n}{6}[\mathbf{C}_\infty^s], \tag{57}$$

$$[\mathbf{K}_{22}^s] = \frac{1}{2}[\mathbf{M}^s] + \frac{3\Delta t_n^2}{24}[\mathbf{K}^s] + \frac{\alpha \Delta t_n}{3}[\mathbf{M}^s] + \frac{\beta \Delta t_n}{3}[\mathbf{K}^s] + \frac{\Delta t_n}{3}[\mathbf{C}_\infty^s], \tag{58}$$

in which $[\mathbf{M}^s]$, and $[\mathbf{K}]^s$ are the mass and stiffness matrix for the solid domain [12], α and β are the coefficients of Rayleigh damping, and matrix $[\mathbf{C}_\infty^s]$ is due to the dashpots placed at truncated boundaries of soil domain which has the form,

$$[\mathbf{C}_\infty^s] = [\mathbf{C}_v^s] + [\mathbf{C}_h^s] = \begin{bmatrix} \mathbf{c}_{11}^v & \mathbf{0} \\ \mathbf{0} & \mathbf{c}_{22}^v \end{bmatrix} + \begin{bmatrix} \mathbf{c}_{11}^h & \mathbf{0} \\ 0 & \mathbf{c}_{22}^h \end{bmatrix} \tag{59}$$

$$[\mathbf{c}_{ii}^v] = \int_{\Gamma_\infty^L \cup \Gamma_\infty^R} c_{ii}^v \mathbf{N}^T \mathbf{N} ds \quad i = 1, 2(\text{no sum}), \tag{60}$$

$$[\mathbf{c}_{ii}^h] = \int_{\Gamma_\infty^B} c_{ii}^h \mathbf{N}^T \mathbf{N} ds \quad i = 1, 2(\text{no sum}), \tag{61}$$

the solid-fluid coupling matrix,

$$[\mathbf{H}^s] = \int_{\Gamma_{fs}^s \cup \Gamma_{fd}^s} \mathbf{N}^T \mathbf{N} \mathbf{n}^s \, ds, \tag{62}$$

and right hand side spatial nodal vectors is given by,

$$\begin{aligned} \{\mathbf{J}_1^s\} = {} & [\mathbf{M}^s]\{\mathbf{V}_0\} - \frac{\Delta t_n}{2}[\mathbf{K}^s]\{\mathbf{U}_0\} + \frac{2\Delta t_n}{3}[\mathbf{C}_h^s]\{\mathbf{V}_1^{in}\} + \frac{2\Delta t_n}{6}[\mathbf{C}_h^s]\{\mathbf{V}_2^{in}\} \\ & + \frac{\Delta t_n}{3}[\mathbf{C}_v^s]\{\mathbf{V}_1^\infty\} + \frac{\Delta t_n}{6}[\mathbf{C}_v^s]\{\mathbf{V}_2^\infty\} + \frac{\Delta t_n}{2}\{\mathbf{F}_1^{ext}\} + \frac{\Delta t_n}{2}\{\mathbf{F}_1^\infty\} \\ & - \frac{\Delta t_n}{2}[\mathbf{H}^s](\{\mathbf{P_h}\} + \{\mathbf{P_0}\}), \end{aligned} \tag{63}$$

$$\{\mathbf{J}_2^s\} = -\frac{\Delta t_n}{2}[\mathbf{K}^s]\{\mathbf{U}_0\} + \frac{2\Delta t_n}{6}[\mathbf{C}_h^s]\{\mathbf{V}_1^{in}\} + \frac{2\Delta t_n}{3}[\mathbf{C}_h^s]\{\mathbf{V}_2^{in}\} + \frac{\Delta t_n}{6}[\mathbf{C}_v^s]\{\mathbf{V}_1^{\infty}\}$$
$$+ \frac{\Delta t_n}{3}[\mathbf{C}_v^s]\{\mathbf{V}_2^{\infty}\} + \frac{\Delta t_n}{2}\{\mathbf{F}_2^{ext}\} + \frac{\Delta t_n}{2}\{\mathbf{F}_2^{\infty}\} - \frac{\Delta t_n}{2}[\mathbf{H}^s](\{\mathbf{P_h}\} + \{\mathbf{P_0}\}), \tag{64}$$

where $\{\mathbf{U}_0\}$, $\{\mathbf{V}_0\}$, and $\{\mathbf{P}_0\}$ are spatial nodal values of $\mathbf{u}$, $\mathbf{v}$, and p at time t_n, $\{\mathbf{P}_h\}$ is nodal values of hydrostatic pressure, and $\{\mathbf{F}_1^{ext}\}$ and $\{\mathbf{F}_2^{ext}\}$ are nodal force vector due to external body forces at time t_n and t_{n+1}, respectively,

$$\{\mathbf{F}_a^{ext}\} = \int_{-1}^{+1}\int_{\Omega_h^s} T_a \mathbf{N} \rho^s \mathbf{b}\, d\Omega d\theta. \tag{65}$$

Further, the force vector $\{\mathbf{F}_{a=1,2}^{\infty}\}$ is due to the free-field stress at the vertical viscous boundaries of soil domain [13], and described by,

$$\{\mathbf{F}_{a=1,2}^{\infty}\} = \int_{-1}^{+1}\int_{\Gamma_\infty^L \cup \Gamma_\infty^R} T_a \mathbf{N} \boldsymbol{\sigma}^{\infty} \cdot \mathbf{n}^s\, ds d\theta. \tag{66}$$

Lastly, nodal values of displacement and pressure field at time t_{n+1} are computed by following expression in a post-processing step.

$$\{\mathbf{U}_2\} = \{\mathbf{U}_0\} + \frac{\Delta t_n}{2}(\{\mathbf{V}_1\} + \{\mathbf{V}_2\}) \tag{67}$$

$$\{\mathbf{P}_2\} = \{\mathbf{P}_0\} + \frac{\Delta t_n}{2}(\{\mathbf{Q}_1\} + \{\mathbf{Q}_2\}) \tag{68}$$

6. Numerical examples

In this section, ST/FEM with block iterative algorithm has been employed to study the response of the concrete gravity dam to the horizontal earthquake motion (see [10]). In the numerical modeling two cases are considered; (i) dam-reservoir (DR) system, in which foundation is considered to be rigid, and (ii) dam-reservoir-soil (DRS) system, in which the foundation is an elastic deformable body.

Figure 7 depicts the physical dimensions of the dam-reservoir system. Length of the reservoir in upstream direction is 200 m, and length of the soil domain in horizontal and vertical direction is 440 m and 150 m, respectively. For the dam, elastic modulus, E, mass-density, ρ^s, and Poisson's ratio, ν, are 28.0 GPa, 2347.0 kg/m^3, and 0.20 respectively, and for the foundation, $E = 40.0$ GPa, $\rho = 2551.0$ kg/m^3, and

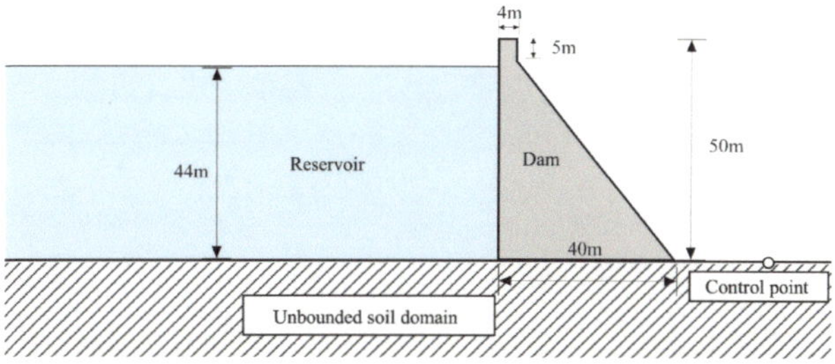

Figure 7.
Physical dimensions of dam-reservoir system (after [10]).

$\nu = 0.20$. Material damping in the solid domain is modeled by Rayleigh damping with $\xi = 5\%$ viscous damping specified for the foundation and dam separately.

Figure 8 represents the accelerogram (horizontal component) recorded at a control point on the free surface; the maximum and minimum values of acceleration are 396.7 Gal (at $t = 15.19$ s) and -449.6 Gal (at $t = 14.82$ s), respectively. Further, numerical simulations are performed for a total time duration of 45 s with a uniform time step size $\Delta t = 0.01$. Time history graphs of acceleration at the crest of the dam in DR and DRS systems are plotted in **Figure 9** where it can be seen that

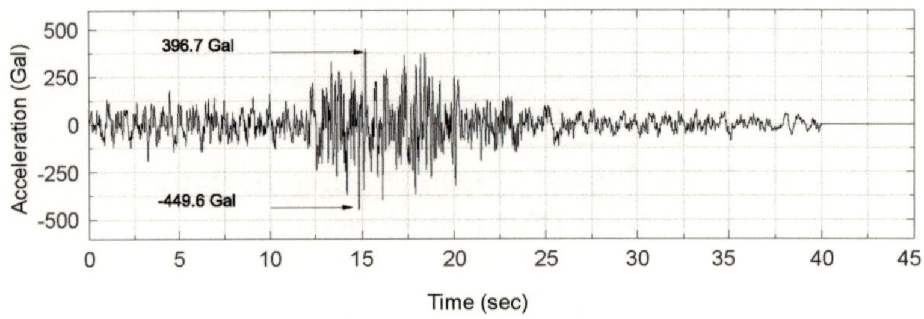

Figure 8.
Time history of horizontal component of ground motion recorded at free-surface.

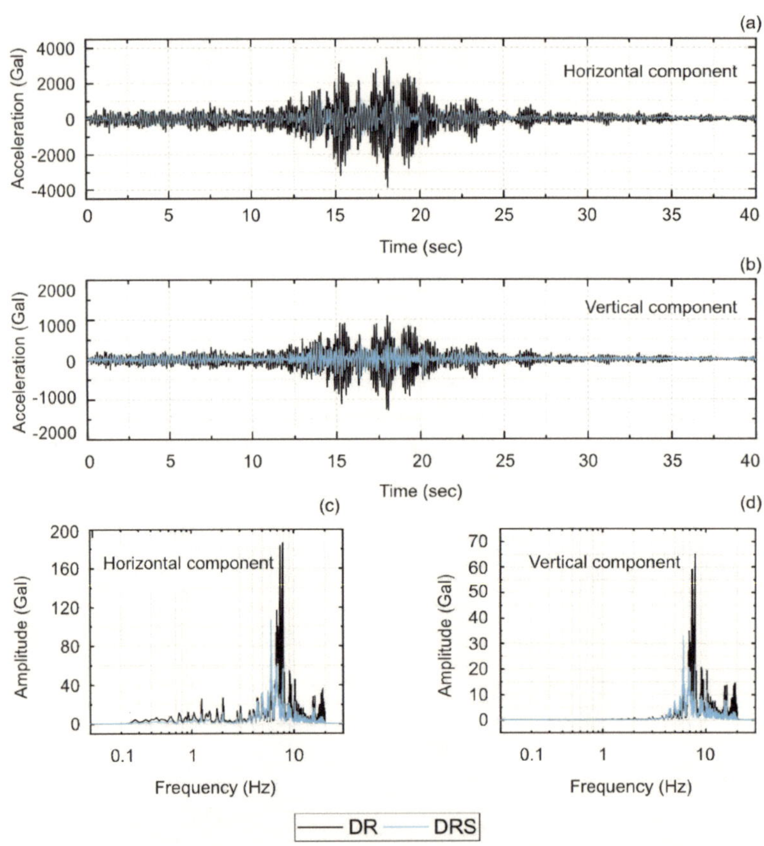

Figure 9.
Acceleration response at the crest of dam in DR and DRS system; time history of (a) horizontal component and (b) vertical component of acceleration, and Fourier spectrum of (c) horizontal component and (d) vertical component of acceleration (after [10]).

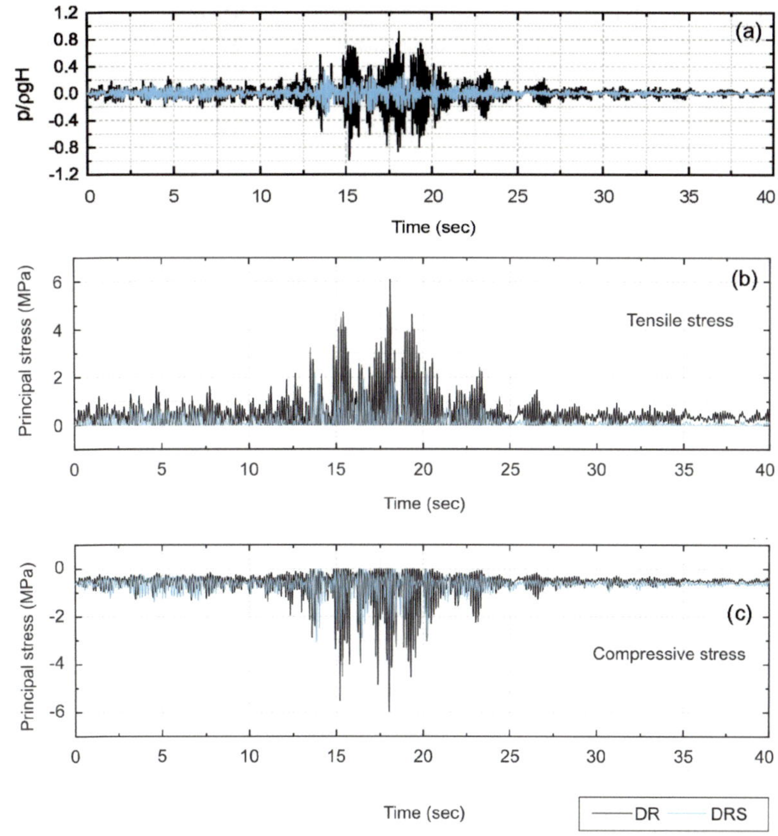

Figure 10.
Temporal response of (a) normalized hydrodynamic pressure, (b) principal tensile stress and (c) principal compressive stress at the base of dam in DR and DRS system (after [10]).

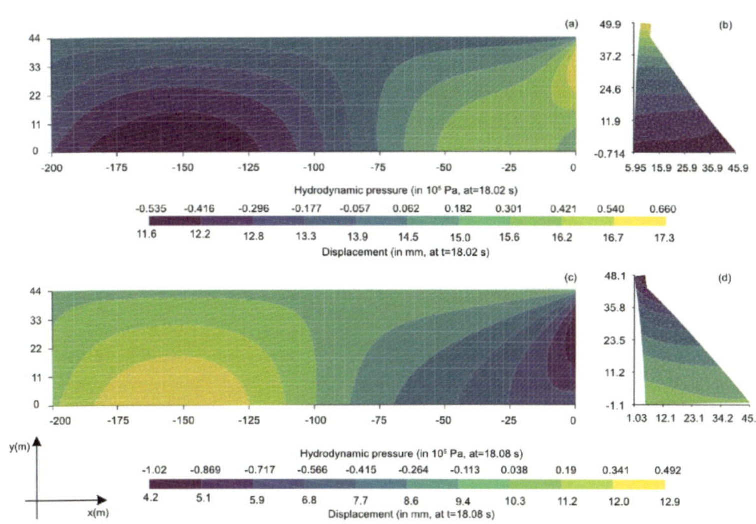

Figure 11.
(a) and (c): Hydrodynamic pressure field in the reservoir, and (b) and (d): magnified deformed configuration of dam at times t = 18:02 s and t = 18:08 s (after [10]).

the deformation characteristics of underlying foundation significantly decreases the responses for dam. To this end, absolute maximum value of horizontal and vertical component of acceleration obtained for DRS are 1489.89 Gal and 597.47 Gal, respectively, and for the DR system these values are equal to 3897.10 Gal and 1274.65 Gal. In addition, Fourier spectrum of time-series of acceleration at the crest indicates that in the case of DRS there is a significant decay in the amplitudes and an elongation of time period as compare to the DR system.

Interestingly, in both cases, it is observed that the critical location for pressure is at the base of the dam. **Figure 10** presents the evolution of p, maximum principle tensile and compressive stresses with time at this location. It is clearly visible that dynamic interactions between the dam-reservoir and the deformable underlying ground significantly lower the hydrodynamic pressure and maximum stresses in the dam. Lastly, the hydrodynamic pressure field and deformed configuration of dam (magnified 500 times) in the DRS system at time t = 18.02 s and t = 18.08 s are presented in **Figure 11**.

7. Conclusions

In this chapter, novel concepts of time-discontinuous Galerkin (tDGM) method is presented. A method called v-tDGM is derived to solve second order ODEs in time. In this method velocity is the primary unknown and it remain discontinuous at discrete times. Thereby, the time-continuity of velocity is satisfied in a weak sense. However, displacement is obtained by time-integration of the velocity in a post-processing step by virtue of which it is continuous in time. It is demonstrated that the present method is unconditionally stable and third-order accurate in time for linear interpolation of velocity in time. Therefore, it can be stated that the numerical characteristics of the v-ST/FEM scheme, therefore, make it highly suitable for computing the response of bodies subjected to dynamic loading conditions, such as fast-moving loads, impulsive loading, and long-duration seismic loading, among others.

Subsequently, ST/FEM is used to compute the response of a dam-reservoir-soil (DRS) system to the earthquake loading while considering all types of dynamic interactions. An auxiliary variable q representing the first order time derivative of the pressure is treated as the primary unknown for the reservoir domain. Similarly, velocity $\mathbf{v}$ is the primary unknown for the solid domain. Both $\mathbf{v}$ and q are interpolated such that they remain discontinuous at the discrete times. Hydrodynamic pressure and displacements are the secondary unknowns in the present formulation which are computed by time integration of q and $\mathbf{v}$, respectively. It is concluded that the dynamic interactions between the dam-reservoir system and the underlying deformable foundation significantly dampen the seismic response of the dam and the reservoir and elongate the time period of the acceleration response of the dam.

Author details

Vikas Sharma[1*], Akira Murakami[2] and Kazunori Fujisawa[2]

1 Indian Institute of Technology Bombay, Mumbai, India

2 Kyoto University, Kyoto, Japan

*Address all correspondence to: vickysharma0812@gmail.com

References

[1] Hughes TJR. Analysis of transient algorithms with particular reference to stability behavior. In: Computational Methods in Mechanics. Vol. 1. Amsterdam, New York: North-Holland Publishing Co.; 1983. pp. 67-155

[2] Hughes TJR, Hulbert GM. Space-time finite element methods for elastodynamics: Formulations and error estimates. Computer Methods in Applied Mechanics and Engineering. 1988;**66**(3):339-363

[3] Li XD, Wiberg N-E. Structural dynamic analysis by a time-discontinuous galerkin finite element method. International Journal for Numerical Methods in Engineering. 1996;**39**(12):2131-2152

[4] Sharma V, Fujisawa K, Murakami A. Velocity-based time-discontinuous galerkin space-time finite element method for elastodynamics. Soils and Foundations. 2018;**58**(2):491-510

[5] Sharma V, Fujisawa K, Murakami A. Space–time fem with block-iterative algorithm for nonlinear dynamic fracture analysis of concrete gravity dam. Soil Dynamics and Earthquake Engineering. 2020;**131**:105995

[6] Bajer CI, Dyniewicz B. Virtual functions of the space–time finite element method in moving mass problems. Computers & Structures. 2009;**87**(7–8):444-455

[7] Bause M, Radu FA, Köcher U. Space–time finite element approximation of the biot poroelasticity system with iterative coupling. Computer Methods in Applied Mechanics and Engineering. 2017;**320**:745-768

[8] Bonelli A, Bursi OS, Mancuso M. Explicit predictor–multicorrector time discontinuous galerkin methods for non-linear dynamics. Journal of Sound and Vibration. 2002;**256**(4):695-724

[9] Mancuso M, Ubertini F. An efficient time discontinuous galerkin procedure for non-linear structural dynamics. Computer Methods in Applied Mechanics and Engineering. 2006;**195**(44–47):6391-6406

[10] Sharma V, Fujisawa K, Murakami A. Space-time finite element procedure with block-iterative algorithm for dam-reservoir-soil interaction during earthquake loading. International Journal for Numerical Methods in Engineering. 2019;**120**(3):263-282

[11] Hughes TJR. The Finite Element Method: Linear Static and Dynamic Finite Element Analysis. Mineola, New York: Dover Publications; 2000

[12] Zienkiewicz OC, Taylor RL. The Finite Element Method for Solid and Structural Mechanics. Butterworth-Heinemann; 2014

[13] Løkke A, Chopra AK. Direct finite element method for nonlinear analysis of semi-unbounded dam–water–foundation rock systems. Earthquake Engineering and Structural Dynamics. 2017;**46**(8):1267-1285

Chapter 4

Long-Term Behavior of Coarse-Grained Rockfill Material and Their Constitutive Modeling

Erich Bauer

Abstract

For the long-term behavior and safety assessment of rockfill dams, not only the shape of the dam body, the loading history, the geological condition of the dam foundation and abutments, the assessment of possible seismic hazards and seepage events caused by defects of the sealing are important, but also the time dependent mechanical behavior of the dam materials used can be of significant influence. In this paper a novel hypoplastic constitutive model for moisture sensitive, coarse-grained rockfill materials is presented. In the constitutive equations, the so-called solid hardness is a key parameter to reflect the influence of the state of weathering on the mechanical response. With respect to the evolution equation for the solid hardness, creep and stress relaxation can be modeled for dry and wet states of the material in a unified manner. The performance of the model is demonstrated by comparing the numerical simulation with experimental data.

Keywords: rockfill material, wetting behavior, creep, stress relaxation, hypoplasticity

1. Introduction

Rockfill dams have become very popular among dam engineers due to of their simple construction sequence, short construction period and low costs compared to concrete dams. An economic aspect also lies in the fact that rockfill materials of different types are usually available on site can be used in appropriate zones of the dam. The frequent use of weathered and moisture-sensitive rockfill materials requires the precise recognition of the mechanical behavior of these materials under the expected load and environmental conditions. The prediction of the post-construction settlements of rockfill dams is a challenging task because of uncertainties of environmental events, which can occur during the entire lifetime of the dam [1, 2]. Changes of the moisture content of weathered rockfill material can lead to sudden settlements that could have relatively large values, without any changes in the applied load [3–8]. For the long-term behavior of rockfill dams, not only the shape of the dam body, the loading history, the geological condition of the dam foundation and abutments, the assessment of possible seismic hazards and seepage events caused by defects of the sealing are important, but also the time dependent behavior of the dam materials used can be of significant influence [9–11]. Thus, the proper modeling of the time dependent behavior of rockfill material under different

IntechOpen

loading and environmental condition plays an important role for the design, construction, operation and safety assessment of rockfill dams. For instance, post construction settlements may affect the amount of the bending and the crack propagation in concrete slaps of concrete face rockfill dams [12–16].

In engineering disciplines time dependent deformations under constant stress are termed creep, however, this term does not reflect the individual mechanisms that different types of materials can exhibit. Rheological properties of weathered rockfill materials are strongly influenced by the state of weathering and the mechanical and environmental boundary conditions to which the material is subjected. Thus, the concept for the experimental investigations requires an appropriate adaptation to the conditions relevant at the construction site. In this context also scale effects resulting from the differences between the grain size distribution and pre-compaction in laboratory tests and in the filed must be taken into account [17–19].

Long-term deformations are usually an accumulation of deformations related to various events and the proper interpretation of the relevant physical and hydro-chemical mechanisms is important for the evaluation of the data obtained from laboratory experiments and field measurements as well as for the numerical modeling. A general distinction can be made between time independent deformations, which are the instantaneous part of the deformation due to applied load changes, and time dependent deformations, which can also take place under constant load. The former, for instance, can be initiated by rapid changes of the water level in the reservoir, the change of the effective stresses caused by a change of suction of fine grained materials, hydraulic fracturing and piping as a result of the seepage-driven internal erosion of solid particles. Time dependent deformations are influenced by the mineralogical composition of the solid material, the frequency and the orientation of micro-cracks, the grain size distribution, the pre-compaction, the moisture content, the stress state and the evolution of weathering [20–24]. Progressive weathering caused by mechanical and hydro-chemical weathering has a significant influence on the time dependent process of the degradation of the solid hardness and as a consequence on the resistance of the material against compaction and shearing. In dam engineering it is common to differentiate between long-term creep and so-called collapse settlements. While the former is related to the rheological properties of the rockfill material and controlled by gravity load and the effect of water impounding, the latter can usually be observed immediately after a change of the moisture content in the stressed rockfill material. Collapse settlements are also called instantaneous wetting deformation and characterized by a spontaneous increase of the deformation velocity.

Changes of the moisture content can be caused by different events like climate changes, leakage as an effect of defects of the dam sealing and the dam foundation, as well as by rainwater infiltration into the dam body. According to Terzaghi [25] collapse phenomenon are also time dependent and related to the decrease in the grain crushing strength, especially at the contact points. Plastification of grain contacts and grain crushing bring about local instabilities in the grain skeleton. Rheological properties of weathered rockfill materials are more pronounced in the wet state of the material than in the dry state [4, 9, 11, 26]. Depending on the state of weathering of the rockfill material a change of the moisture content can initiate an acceleration of the crack propagation of stressed rockfill grains, which leads to a reduction of the solid hardness and consequently to a sudden increase of the settlement velocity. **Figure 1a** shows an example for an increase of the settlement velocity after flooding under a constant vertical stress of -0.8 MPa in an oedometer device [3]. The course of the deformations is qualitatively similar for greywacke and sandstone and can be divided into three parts. After applying the vertical stress on

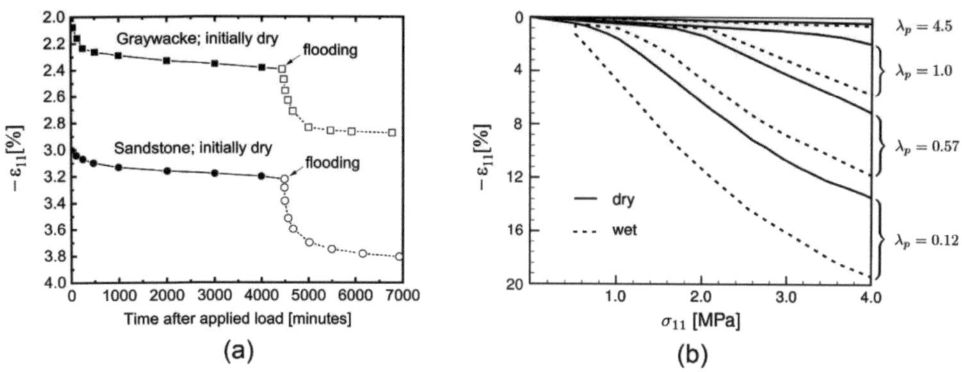

Figure 1.
(a) Time dependent behavior of two different rockfill materials under a constant vertical stress of −0.8 MPa in an oedometer device carried out by Sowers et al. [3]; (b) oedometric compression behavior of broken granite under dry and water saturated conditions and for different pre-compactions [27, 28] (herein λ_p denotes a measure of the relative pre-compaction energy).

the initially dry material, an instantaneous settlement can be detected, which is larger for sandstone. Under constant stress, creep can be observed in the dry state and also following a sudden flooding of the specimen. Flooding leads to a sudden jump of the settlement rate but there is no clear sharp jump in the settlement which indicates that so-called collapse settlements can also assumed to be time dependent. While for the dry material the creep velocity decreases slowly, the high settlement rate immediately after flooding fades out very fast. It is experimentally evident that the compressibility of weathered rockfill material strongly depends on the pre-compaction and moisture content of the material, i.e. the compressibility is higher for a less compacted material and higher for the wet than for the dry material. This is also clearly visible for example from the results of oedoemeter compression tests with weathered broken granite in **Figure 1b** [27, 28]. Great efforts have been made to investigate and to model the complex mechanisms of wetting deformations at the micro- and macro-level, e.g. [29–46].

The focus of this paper is on the time dependent process of degradation of the stiffness of weathered, coarse grained and moisture sensitive rockfill materials and its constitutive modeling under dry and wet states. A particular version of a hypoplastic model developed within the past decade is presented and its performance is verified by comparison of numerical simulations with experimental data. It is an aim of the present paper to propose simple calibration procedures for the material parameters based on standard laboratory tests. As a measure of the state of weathering the so-called "solid hardness of the grain assembly" is a key parameter in the material model proposed [47–52]. Particular attention is paid on a refined modeling of the influence of the coupled behavior between the state of weathering, the stress state and the packing density of the rockfill material on the calibration of those material parameters relevant for collapse settlements, long-time creep and stress relaxation. In this context the modeling of the change of the rheological material parameters during the lifetime of the dam, the concept of the solid hardness is put forward to simulate a repeated acceleration of the degradation of the solid hardness. Such events are relevant for repeated changes in the moisture content of the rockfill material caused by local defects of the sealing and heavy rain water infiltration into the rockfill material of the dam body. While the focus of the present paper is on the theory of constitutive modeling and its calibration, the application of the proposed material model to different types of rockfill dams can be found for instance in [53–56].

The present paper is organized as follows:

In Section 2 resent developments of modeling the compression behavior of weathered, creep and moisture sensitive rockfill materials under dry and wet condition are summarized. To this end the so-called "solid hardness" is defined in the sense of a continuum description and is a state parameter in Bauer's compression law [47, 57, 58]. As an example of how the concept of the solid hardness can be incorporated into a general three dimensional material model the adaptation of the compression law to the bulk modulus of a nonlinear elastic material model is outlined. In incrementally non-linear material models, for instance, the concept of the solid hardness was embedded into non-polar constitutive models, e.g. [58–60] and into micro-polar models, e.g. [61–68]. In the present paper the proposed enhanced constitutive model for the solid hardness is a time dependent quantity and a measure of the state of weathering of the rockfill material [47]. With respect to the evolution equation of the time dependent process of degradation of the solid hardness creep and stress relaxation are modeled in a unified manner. Particular attention is paid to the adaptation of the velocity parameter in the evolution equation of the solid hardness to experimental creep and stress relaxation curves.

Section 3 gives a brief introduction to the hypoplastic constitutive model by Gudehus [69] and Bauer [58] which was originally developed for a constant solid hardness and a time independent material behavior. The incrementally nonlinear constitutive equation describes the stress rate as a function of the current void ratio, the Cauchy stress and the strain rate. In contrast to the concept of elasto-plasticity, the framework of hypoplasticity does not need to decompose the deformation into elastic and plastic parts, which allows a relatively easy calibration of the material parameters involved. The hypoplastic constitutive model captures the extended theory of "critical state soil mechanics" and describes the influence of pressure and density on the incremental stiffness, peak friction angle and dilatancy angle using only eight material parameters. It is also outlined in detail how the compression law by Bauer can be embedded into the hypoplastic model with help of a consistency condition.

In Section 4 the hypoplastic constitutive model shown in Section 3 is extended to describe also time dependent material properties which are relevant for weathered and moisture sensitive rockfill materials. In addition to the current void ratio and stress, the solid hardness and its rate are also state quantities of the extended model. As a consequence creep and stress relaxation properties are usually coupled and the incremental stiffness, peak friction angle, dilatancy angle are also influenced by the evolution of the degradation of the solid hardness.

In Section 5 the capability of the proposed hypoplastic model to model the mechanical behavior of rockfill materials under dry and wet conditions are verified by comparison of numerical simulations with experimental results. In order to simulation sudden changes of the creep velocity initiated by repeated acceleration of the degradation of the solid hardness during the lifetime of the dam an extended version for the evolution equation of the solid hardness is proposed. The enhanced version permits the simulation of multistep degradation of the solid hardness and a refined simulation of collapse settlements and long-time creep.

Throughout the paper the sign convention of rational solid mechanics is adopted, i.e. compressive stresses and strains, and their rates are negative. Indices on vector and tensor components refer to an orthonormal Cartesian basis and the symbol δ_{ij} denotes the Kronecker delta. The summation convention by Einstein over repeated indices is employed. A superimposed dot indicates the material time derivative, e.g. $\dot{e} = \mathrm{d}e/\mathrm{d}t$. All stresses are effective stresses. Other effects of partly saturated materials such as inter-particle capillary forces are negligible for coarse grained rockfill materials.

2. Compression law by Bauer

This section deals with the modeling of essential mechanical properties of weathered and moisture sensitive rockfill materials under isotropic and oedometric compression. To this end the so-called "solid hardness" is defined in the sense of a continuum description and it is a state parameter in the compression law by Bauer. An evolution equation for the degradation of the solid hardness of the rockfill material is used to model the influence of the reduction of the incremental stiffness caused by progressive weathering, instantaneous wetting deformation and long-time creep. Examples for the adaptation of the velocity parameter in the evolution equation of the solid hardness to experimental creep and stress relaxation curves are outlined in detail.

2.1 Introduction of the solid hardness in the sense of a continuum description

Isotropic or oedometric compression tests carried out with various granular materials show qualitatively similar compression curves. At lower stresses the reduction of the void ratio is explained by sliding of neighboring grains against each other and a reorientation of the grain skeleton into a denser state. Under higher pressures the additional compaction is mainly related to progressive grain crushing. In a semi-logarithmic representation, the compression curves show an S-shape which is also clearly visible from the experimental data for two different sand materials in **Figure 2** [70]. For different initial void ratios, the distance between the compression curves becomes smaller with an increase in the mean pressure and for higher pressures the curves merge together. This means that the memory of the material on the initial density fades out due to both grain crushing and reorientation of grains. It is known from experiments that the point of inflection is related to the pressure level where grain crushing becomes dominant. Such a compression behavior was also observed for arbitrary granular materials and it is also verified by

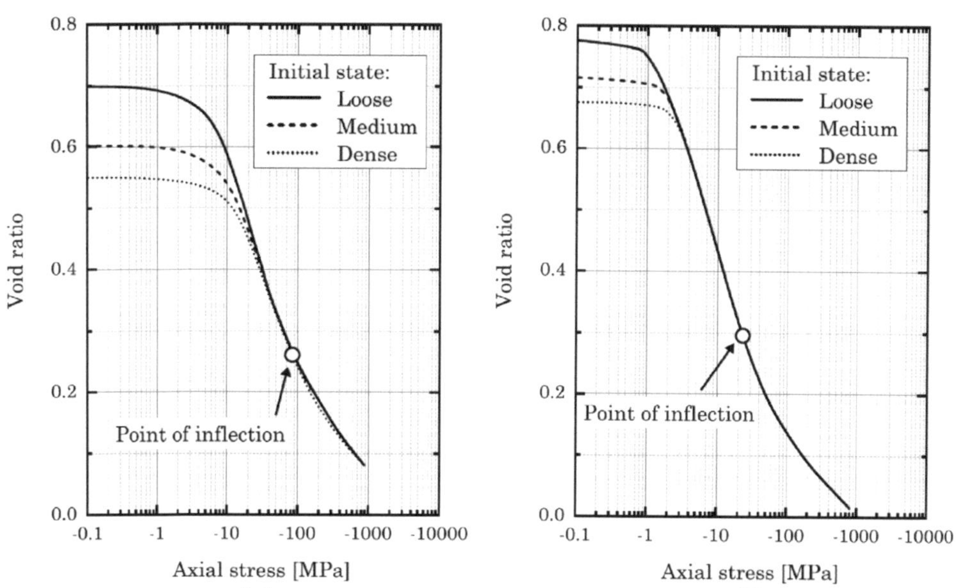

Figure 2.
Decrease of the void ratio with an increase of the axial stress in oedeometer tests on sand specimens starting from different initial void ratios [70]: (a) Cambra sand; (b) Gypsum sand.

numerical simulations with the discrete element method, e.g. [71, 72]. Experimental work has shown that the pressure at the point of inflection depends mainly on the mineral composition and the state of weathering of the solid material. As the point of inflection shows no noticeable influence on the initial density, it is a well-defined state of the material under compression. The pressure where the point of inflection appears is in the following called "solid hardness" and it is a material parameter in the compression law by Bauer [57, 58]. In this context it is worth noting that the solid hardness is defined for the compression behavior of an assembly of grains and does not mean the hardness of a single grain.

In the case of frequently used constitutive models the compression behavior represented in a semi-logarithmic representation is approximated by a straight line as illustrated in **Figure 3a**. However, the approximation is only applicable to a limited pressure range. For higher pressures the compression line, NCL, leads to the non-physical area of negative void ratios. A behavior of this kind cannot occur when using the compression law proposed by Bauer [57, 58]. The exponential function captures the whole pressure range in a consistent manner as depict in **Figure 3b**. In particular, Bauer's isotropic compression law describes the reduction of the void ratio e with an increasing mean effective pressure $p = -(\sigma_{11} + \sigma_{22} + \sigma_{33})/3$ according to the following exponential function:

$$e_i = e_{i0} \exp\left\{-\left(\frac{3p}{h_s}\right)^n\right\}. \tag{1}$$

Eq.(1) represents the upper bound of the pressure dependent maximum void ratio e_i and involves three material parameters. In particular, e_{i0} denotes the maximum void ratio at the initially nearly stress free state, the solid hardness h_s is the value of $3p$ where the compression curve in a semi-logarithmic representation shows the point of inflection, and n is related to the inclination of the compression curve in the point of inflection. The compression law is consistent within the whole pressure range and thus it also takes into account the influence of grain crushing at higher pressures [67].

For rockfill materials the value of the solid hardness is rather high, and usually it cannot be achieved with standard isotropic compression apparatus available in a soil mechanics laboratory. On the other hand, in an oedometer device higher pressures are easier to carry out. Investigations show that for practical application the solid hardness can also be calibrated with sufficient accuracy if data are used which are

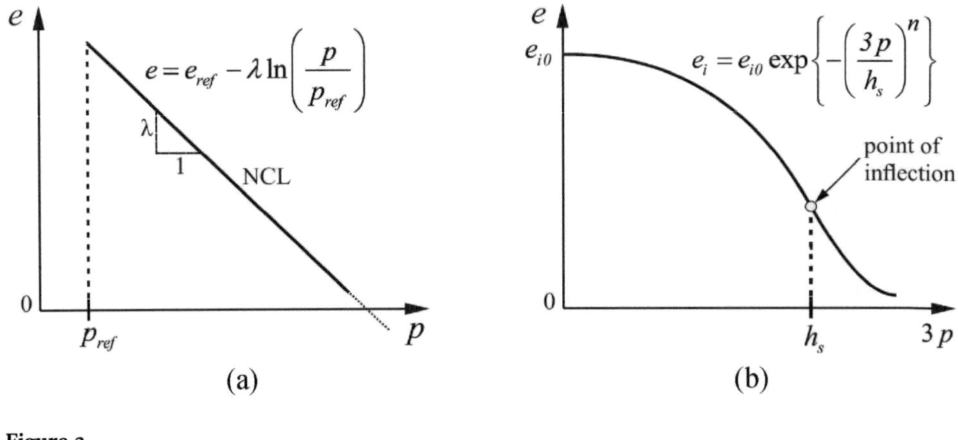

Figure 3.
Illustration of compression laws in the semi-logarithmic representation $e - \ln(3p)$*: (a) normal compression line (NCL); (b) compression law by Bauer [58].*

obtained from an oedometer test instead of an isotropic test. Standard oedoemeter devices usually only allow the measuring of the vertical stress σ_V, so that the lateral stress $\sigma_H = K_0\sigma_V$ must be estimated for instance with the help of the Jaky formula [73], i.e. for the coefficient of the earth pressure at rest $K_0 \approx 1 - \sin\varphi_c$. Herein φ_c denotes the critical friction angle. Then the mean pressure can be calculated form $p = -\sigma_V(1 + 2K_0)/3$.

The embedding of the compression law (1) into general 3-D constitutive models can be accomplished with the help of a consistency condition. As a heuristic example the embedding into an elastic material model is demonstrated in Appendix A. The implementation of Eq.(1) into an enhanced constitutive model for rockfill materials is discussed in Section 3 and outlined in detail in Appendix B.

2.2 Time dependent process of degradation of the solid hardness

The degradation of the solid hardness caused by progressive weathering of the rockfill material is a time dependent process and causes a reduction of resistance of the material against shearing and compaction. A chemical reaction of the weathered rockfill material with water can accelerate the process of weathering leading to grain breakage and as a consequence to collapse settlements and creep deformations. Experimental investigations show that the solid hardness of the dry material is higher than the solid hardness of the wet material, and that the transition from the dry to the wet state is a time dependent process as illustrated in **Figure 4**.

In order to take the current state of weathering into account, the constant solid hardness h_s in Eq.(1) is replaced by the state quantity h_{st}, i.e.

$$e_i(p,\ h_{st}) = e_{i0} \exp\left\{-\left(\frac{3p}{h_{st}}\right)^n\right\}. \tag{2}$$

The change of the void ratio with time t, i.e. $\mathrm{d}e/\mathrm{d}t = \dot{e} < 0$, then depends on the current state of e, p, h_{st}, and the rate quantities $\dot{p}$ and $\dot{h}_{st}$, i.e.

$$\dot{e}\left(e,\ p,\ h_{st},\ \dot{p},\ \dot{h}_{st}\right) = n\, e \left(\frac{3p}{h_{st}}\right)^n \left(\frac{\dot{h}_{st}}{h_{st}} - \frac{\dot{p}}{p}\right). \tag{3}$$

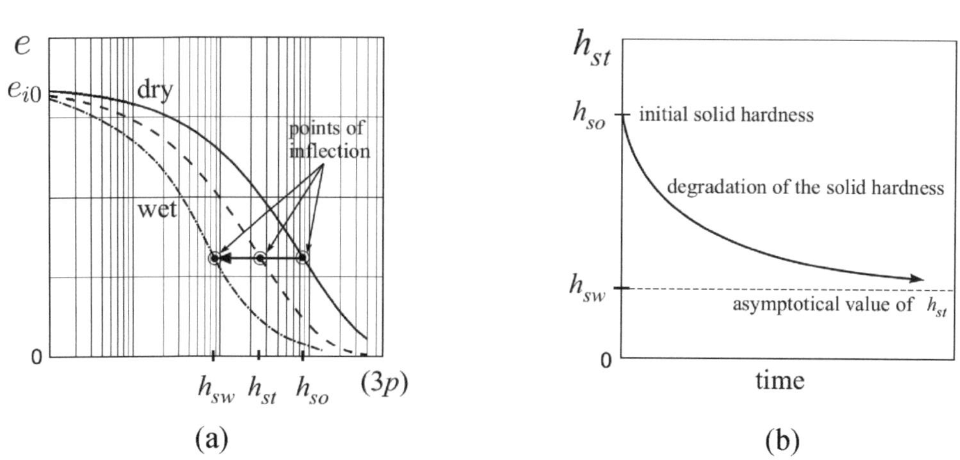

Figure 4.
(a) Transition of the compression curve from the dry to the wet state; (b) time dependent degradation of the solid hardness.

Substituting the identity $\dot{e} = (1+e)\,\dot{\varepsilon}_V$ into Eq.(3) the following relation for the volume strain rate $\dot{\varepsilon}_V$ is obtained:

$$\dot{\varepsilon}_V = \frac{n\,e}{1+e}\left(\frac{3p}{h_{st}}\right)^n\left(\frac{\dot{h}_{st}}{h_{st}} - \frac{\dot{p}}{p}\right). \tag{4}$$

The evolution Eqs. (3) and (4) describe creep and stress relaxation in a unified manner. In particular, for the special case that during degradation of the solid hardness the pressure is kept constant, i.e. $\dot{p} = 0$, the evolution equation for creep can be obtained. For instance, under the constant isotropic stress p_o the evolution equation for the volume strain and for the corresponding void ratio reads:

$$\dot{\varepsilon}_V = \frac{n\,e}{1+e}\left(\frac{3p_o}{h_{st}}\right)^n\left(\frac{\dot{h}_{st}}{h_{st}}\right), \tag{5}$$

and

$$\dot{e} = n\,e\left(\frac{3p_o}{h_{st}}\right)^n\left(\frac{\dot{h}_{st}}{h_{st}}\right), \tag{6}$$

respectively. With respect to the state quantities at the reference time $t = 0$, i.e. $e(t=0) = e_0$ and $h_{st}(t=0) = h_{so}$, the integration of Eq.(6) leads to:

$$e(t) = e_o \exp\left\{\left(\frac{3p_o}{h_{so}}\right)^n - \left(\frac{3p_o}{h_{st}(t)}\right)^n\right\}. \tag{7}$$

The integration of the identity $\dot{e} = (1+e)\,\dot{\varepsilon}_V$ with respect to $\varepsilon_V(t=0) = 0$ and $e(t=0) = e_0$ leads to:

$$e(t) = (1+e_0)\exp\{\varepsilon_V\} - 1. \tag{8}$$

The comparison of Eq.(7) with Eq.(8) yields the creep strain ε_V depending on the current state of the solid hardness h_{st}:

$$\varepsilon_V(t) = \ln\left[\frac{e_0 \exp\left\{(3p_o/h_{so})^n - (3p_o/h_{st}(t))^n\right\} + 1}{(1+e_0)}\right]. \tag{9}$$

On the other hand for the special case that the volume is kept constant, i.e. $\dot{\varepsilon}_V = 0$, Eq.(4) leads the evolution equation for stress relaxation:

$$\dot{p} = p\left(\frac{\dot{h}_{st}}{h_{st}}\right). \tag{10}$$

With respect to $p(t=0) = p_o$ and $h_{st}(t=0) = h_{so}$ the integration of Eq.(10) leads the stress relaxation law:

$$p(t) = \frac{p_o}{h_{so}}\,h_{st}(t). \tag{11}$$

For the irreversible degradation of the solid hardness with time, i.e. $dh_{st}/dt = \dot{h}_{st} < 0$, an appropriate constitutive relation must be chosen. The constitutive function should capture the influence of the essential state quantities relevant for the complex chemical and mechanical response of the stressed material under different environmental events, such as the repeated change of the moisture content caused by rain water infiltration into the dam body. On the other hand, the evolution equation should allow an easy calibration based on the available experimental data. Thus, a compromise must usually be found for practical applications. In this paper the following phenomenological evolution equation based on data from creep tests or stress relaxation tests is considered [67]:

$$\dot{h}_{st} = -\frac{1}{c}(h_{st} - h_{sw}), \tag{12}$$

where parameter c has the dimension of time and parameter h_{sw} denotes the final value of the degraded solid hardness. The integration of Eq.(12) yields for the solid hardness depending on the time t:

$$h_{st}(t) = h_{sw} + (h_{so} - h_{sw}) \exp\left\{-\frac{t}{c}\right\}, \tag{13}$$

where parameter h_{so} denotes the value of the solid hardness at the reference time $t = 0$ and it can be obtained by adaptation of Eq.(1) to the compression curve of the material in the dry state. Moisture sensitive and weathered rockfill materials usually exhibits a higher compressibility and thus the solid hardness is lower than for the dry material. The value of parameter h_{sw} denotes the asymptotic value for $t \to \infty$ and can be adapted to the compression curve for the material under wet condition. The degraded solid hardness, h_{sw}, takes into account contributions of both creep deformations and collapse settlements. For a refined modeling the evolution equation for the degradation can be extended to distinguish collapse deformation and long-term deformation as demonstrated in Section 5. Parameter c has the dimension of time and can be calibrated from the volume creep curve or from the stress relaxation test as outlined in Appendix B. When experimental data are only available for creep curves under deviatoric stresses the mechanical response is also affected by the current packing density and stress deviator so that parameter h_{sw} and c must be calibrated using the coupled constitutive model outlined in Section 4. It is also worth noting that parameter h_{sw} and c are usually not material constants and may change as a consequence of repeated changes of environmental conditions during the whole lifetime of the dam as also discussed in Section 5.2.

Weathered rockfill materials can also undergo creep deformation and stress relaxation under dry condition and it can be distinguished between instantaneous deformation and time dependent deformation. The time dependent compressibility of the dry material can be explained by delayed grain crushing and it can also be modeled by a degradation of the solid hardness. In this case h_{sw} in Eq.(12) denotes the asymptotical value of h_{st} related to the dry state of the material.

The value of parameter h_{sw} can also be back calculated from the asymptotic value of the creep curve or stress relaxation curve. In particular, the asymptotic value of the volume strain can be estimated form the creep curve, i.e. $\varepsilon_{V\infty} = \varepsilon_V(t \to \infty)$, so that the quantity of $h_{sw} = h_{st}(t \to \infty)$ can be calculate from Eq.(9) for $t \to \infty$, i.e.

$$h_{sw} = 3p_o\left[\left(\frac{3p_o}{h_{so}}\right)^n - \ln\left[\frac{(1+e_0)}{e_0}\exp\{\varepsilon_{V\infty}\} - 1\right]\right]^{(-1/n)}. \tag{14}$$

From a pure stress relaxation test the value of h_{sw} can also be calculated from Eq.(11) to:

$$h_{sw} = h_{so} \frac{p_\infty}{p_o}, \tag{15}$$

where $p_\infty = p(t \to \infty)$ denotes value of the mean stress at the end of stress relaxation. The value can be estimated from the asymptote of the stress relaxation curve.

3. Embedding the solid hardness into hypoplasticity

To demonstrate the embedding of the solid hardness h_o into a 3-D constitutive model the hypoplastic model for cohesionless soil proposed by Gudehus [69] and Bauer [58] is considered in the following. In particular, the components of the objective stress tensor for the general form of the nonlinear constitutive equation reads:

$$\sigma_{ij}^{\nabla} = f_s \left[\hat{a}^2 \dot{\varepsilon}_{ij} + (\hat{\sigma}_{kl}\, \dot{\varepsilon}_{kl})\, \hat{\sigma}_{ij} + f_d \hat{a} \left(\hat{\sigma}_{ij} + \hat{\sigma}_{ij}^* \right) \sqrt{\dot{\varepsilon}_{kl}\, \dot{\varepsilon}_{kl}} \right]. \tag{16}$$

The quantities in Eq.(16) denote:

σ_{ij}^{∇} ... objective stress rate
$\dot{\varepsilon}_{ij}$ ... strain rate
σ_{ij} ... effective Cauchy stress
$\hat{\sigma}_{ij}$ ... normalized effective Cauchy stress, i.e. $\hat{\sigma}_{ij} = \sigma_{ij}/\sigma_{kk}$
$\hat{\sigma}_{ij}^*$ ... deviatoric part of $\hat{\sigma}_{ij}$, i.e. $\hat{\sigma}_{ij}^* = \hat{\sigma}_{ij} - \delta_{ij}/3$
$\hat{a}$ ... critical stress factor
f_s ... stiffness factor
f_d ... relative density factor

Because of the nonlinearity of $\sqrt{\dot{\varepsilon}_{kl}\, \dot{\varepsilon}_{kl}}$ the hypoplastic constitutive Eq.(1) is apt to models irreversible deformations without any decomposition of the deformation into elastic and plastic parts. Factor $\hat{a}$ is a function of the critical friction angle, φ_c, and the invariants of the stress deviator. The adaptation of $\hat{a}$ to the stress limit condition by Matsuoka and Nakai [74] is illustrated in **Figure 5a** and represented in the hypoplastic model as [67]:

$$\hat{a} = \frac{\sin \varphi_c}{3 - \sin \varphi_c} \left[\sqrt{\frac{8 - 9\left(\hat{\sigma}_{pq}^* \hat{\sigma}_{pq}^* + \hat{\sigma}_{kl}^* \hat{\sigma}_{lm}^* \hat{\sigma}_{mk}^* \right)}{3 - 9\left(\hat{\sigma}_{kl}^* \hat{\sigma}_{lm}^* \hat{\sigma}_{mk}^* \right) / \left(\hat{\sigma}_{pq}^* \hat{\sigma}_{pq}^* \right)}} - \sqrt{\hat{\sigma}_{kl}^* \hat{\sigma}_{kl}^*} \right]. \tag{17}$$

The peak friction angle and dilatancy are predictions of the constitutive Eq. (16) and not material constants. The influence of pressure and density on the peak friction angle and dilatancy angle is modeled using the pressure dependent relative density factor f_d, which represents a relation between the current void ratio e, the critical void ratio, e_c and the minimum one e_d, i.e.

$$f_d = \left(\frac{e - e_d}{e_c - e_d} \right)^{\alpha}, \tag{18}$$

where α is a material parameter. The change of the volume of the rockfill material is related to a change of the void ratio which can be expressed by the following evolution equation:

$$\dot{e} = (1+e)\dot{\varepsilon}_V. \tag{19}$$

In Eq.(18) the pressure dependent quantities e_c and e_d can be related to the compression law (1) according to the postulate by Gudehus [69], i.e.:

$$\frac{e_c}{e_{c0}} = \frac{e_d}{e_{d0}} = \frac{e_i}{e_{i0}} = \exp\left\{-\left(\frac{3p}{h_s}\right)^n\right\}, \tag{20}$$

where e_{c0}, e_{d0} and e_{i0} are the values for $p \simeq 0$. Eq.(20) describes a reduction of the limit void ratios and the critical void ratio with increasing mean stress as illustrated in **Figure 5b**.

The influence of the pressure and density on the incremental stiffness is modeled using the stiffness factor f_s, which can be represented as the product of three parts, i.e.:

$$f_s = f_e f_\sigma f_b. \tag{21}$$

The first term on the right hand side of Eq.(21) is the density dependent part and a relation between the pressure dependent maximum void ratio e_i and the current void ratio e, i.e.:

$$f_e = \left(\frac{e_i}{e}\right)^\beta, \tag{22}$$

where β is a material parameter. Factor f_σ takes into account a decrease in the incremental stiffness with an increase of $\hat{\sigma}_{kl}\hat{\sigma}_{kl}$, i.e.:

$$f_\sigma = \frac{1}{\hat{\sigma}_{kl}\hat{\sigma}_{kl}}. \tag{23}$$

Factor f_b is referred to as the barotropy factor and is obtained by a consistency condition, which allows the embedding of the compression law by Bauer into the hypoplastic constitutive Equation [67, 69]. In particular, for monotonic isotropic

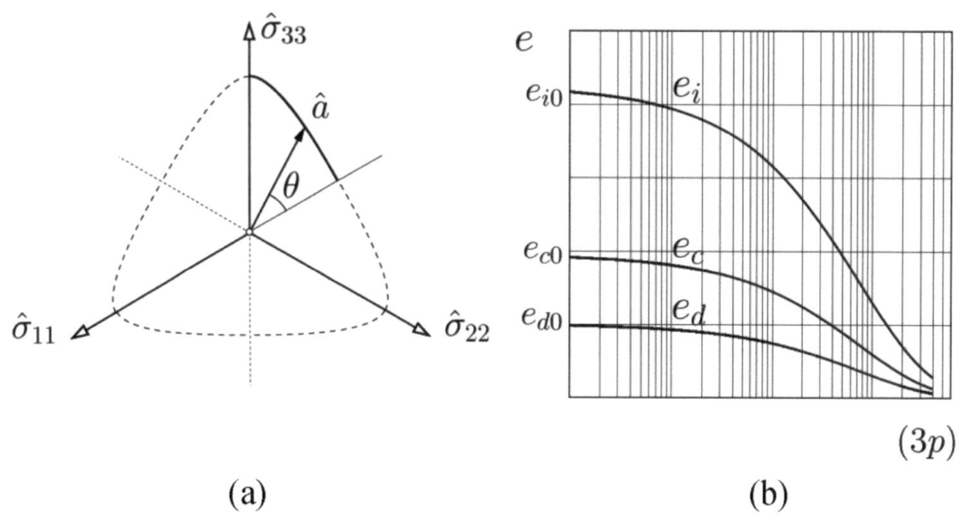

Figure 5.
(a) Stress limit condition by Matsuoka and Nakai [74] in the deviator plane; (b) pressure dependence of the maximum void ratio e_i, critical void ratio e_c and minimum void ratio e_d in the phase diagram of void ratios.

compression the rates of the mean stresses obtained from the compression law (1) and the hypoplastic constitutive Eq. (16) must coincide. The detailed derivation of the consistency condition is outlined in Appendix C. For factor f_b one obtains:

$$f_b = \frac{h_s\,(1+e_i)}{n\,e_i}\left[\frac{8\,\sin^2\varphi}{(3-\,\sin\varphi)^2}+1-\frac{2\sqrt{2}\sin\varphi}{(3-\,\sin\varphi)}\left(\frac{e_{i0}-e_{d0}}{e_{c0}-e_{d0}}\right)^{\alpha}\right]^{-1}\left(\frac{3p}{h_s}\right)^{(1-n)}. \tag{24}$$

The hypoplastic constitutive model also captures properties according to the theory of the "critical state soil mechanics" [75]. In particular, in critical states the pressure dependent void ratio $e = e_c$ and thus, factor $f_d = 1$. By applying the condition that in critical states the stress rate and volume strain rate must simultaneously vanishing under continuous deformation, i.e. $\boldsymbol{\sigma}^{\nabla} = \mathbf{0}$, $\dot{\varepsilon}_V = 0$ and $\dot{\boldsymbol{\varepsilon}} \neq \mathbf{0}$, to Eq.(16) one obtains: $\hat{a}^c = \|\hat{\boldsymbol{\sigma}}^{*c}\| = \sqrt{\hat{\sigma}_{kl}^{*c}\,\hat{\sigma}_{kl}^{*c}}$. Herein $\hat{a}^c$ is the corresponding value of $\hat{a}$ obtained from Eq.(17) and $\|\hat{\boldsymbol{\sigma}}^{*c}\|$ denotes the norm of the normalized deviatoric stress tensor in the critical state [76]. With respect to the critical friction angle φ_c, which is defined for the critical state reached under monotonic triaxial compression, factor $\hat{a}^c$ takes the value: $\hat{a}^c = \sqrt{8/3}\,\sin\varphi_c/(3-\,\sin\varphi_c)$.

The hypoplastic constitutive Eq. (16) contains eight material parameters, i.e. $\varphi_c, h_s, n, e_{io}, e_{co}, e_{do}, \alpha, \beta$, which can be calibrated from standard laboratory tests as outlined in [58]. The capacity of the model in simulating the essential features of an initially loose and dense state of the soil using the same set of material parameters is demonstrated for instance in [58].

4. Extension of the hypoplastic model to weathered and moisture sensitive rockfill materials

Although the hypoplastic constitutive Eq. (16) is of the rate type, the material behavior described is rate independent. In order to take into account rheological properties of weathered and moisture sensitive rockfill materials in the hypoplastic model outlined in the previous section the constant solid hardness h_s is replaced with a state dependent solid hardness $h_{st}(t)$. Furthermore, an additional term is added in the extended hypoplastic model to capture the influence of the rate of the solid hardness, $\dot{h}_{st}(t)$, on the mechanical response. The extended hypoplastic constitutive equation reads [47]:

$$\sigma_{ij}^{\nabla} = f_s\left[\hat{a}^2\dot{\varepsilon}_{ij} + (\hat{\sigma}_{kl}\,\dot{\varepsilon}_{kl})\,\hat{\sigma}_{ij} + f_d\,\hat{a}\left(\hat{\sigma}_{ij}+\hat{\sigma}_{ij}^*\right)\sqrt{\dot{\varepsilon}_{kl}\,\dot{\varepsilon}_{kl}}\right] + \frac{\dot{h}_{st}}{h_{st}}\left[\frac{1}{3}\sigma_{kk}\,\delta_{ij} + \kappa\,\sigma_{ij}^*\right]. \tag{25}$$

In Eq.(25) the degradation of the solid hardness is coupled with the evolution of stress and strain. As the pressure dependent limit void ratios defined in Eq.(20) are also functions of the current state of the solid hardness a degradation of the solid hardness also results in a reduction of the limit void ratios. For a lower solid hardness the stiffness factor f_s and density factor f_d are also lower, so that the incremental stiffness and the peak friction angle predicted with the constitutive Eq. (25) are also lower. In the last term on the right hand side of Eq.(25) parameter κ scales the effect of the stress deviator on stress relaxation and creep. Numerical studies in [53, 48, 49] show that parameter κ of Eq.(25) has a significant influence on the inclination of the volume creep path $\mathrm{d}\varepsilon_V/\mathrm{d}\varepsilon_{11}$. The calibration of parameter κ is discussed in Section 5.

For general boundary conditions the evolution of creep is obtained for $\overset{\nabla}{\sigma}_{ij} = 0$, i.e.:

$$f_s\left[\hat{a}^2\dot{\varepsilon}_{ij} + (\hat{\sigma}_{kl}\,\dot{\varepsilon}_{kl})\,\hat{\sigma}_{ij} + f_d\,\hat{a}\left(\hat{\sigma}_{ij} + \hat{\sigma}^*_{ij}\right)\sqrt{\dot{\varepsilon}_{kl}\,\dot{\varepsilon}_{kl}}\right] + \frac{\dot{h}_{st}}{h_{st}}\left[\frac{1}{3}\sigma_{kk}\,\delta_{ij} + \kappa\sigma^*_{ij}\right] = 0, \quad (26)$$

and for pure stress relaxation, i.e. stress changes under $\dot{\varepsilon}_{ij} = 0$, the constitutive Eq. (25) reduced to:

$$\overset{\nabla}{\sigma}_{ij} = \frac{\dot{h}_{st}}{h_{st}}\left[\frac{1}{3}\sigma_{kk}\,\delta_{ij} + \kappa\sigma^*_{ij}\right]. \quad (27)$$

For the special case of isotropic stress states under $e = e_i$ Eq. (26) turns into the creep law (5) and Eq. (27) ends up with the evolution law (10) for stress relaxation. Parameters c and h_{sw} involved in the evolution Eq. (12) can be calibrated from isotropic compression tests as outlined in Appendix B. It should be noted that a decoupling of creep and stress relaxation is only possible under special boundary conditions. In general boundary value problems creep and stress relaxation are usually coupled. Time dependent deformations and local stress rearrangements can then interact each other, so that in such cases the coupled constitutive Eq. (25) needs to be applied. When the degradation of the solid hardness has been completed, i.e. the final value h_{sw} of the solid hardness is reached, the constitutive Eq. (25) turns into the constitutive Eq. (16).

5. Numerical simulations

In the following the performance of the proposed constitutive concept for modeling the degradation of the solid hardness is demonstrated by comparing experimental data with the results of numerical simulations of a multistep, one dimensional creep test, and triaxial creep tests under different deviatoric stresses. For numerical modeling of wetting induced settlements a smooth transition from an almost sudden settlement, i.e. collapse settlement, to the creep deformation is assumed. A smooth transition of the kind is also observed in laboratory experiments and can be modeled with an extended evolution equation of the solid hardness as demonstrated in the following.

5.1 Combined modeling of collapse deformation and creep

In this subsection the results of wetting experiments with broken sandstone carried out by Fu et al. [77] are considered for the simulation with a simplified version of the hypoplastic constitutive model proposed in [60], which differs from the hypoplastic model (25) in that for the density factor the simplified relation $f_d = (e/e_c)^\alpha$, for $\beta = 1$, and for the critical void ratio curve the relation $e_c = e_{co}\exp\{-(3p/73.5)^n\}$ is used. In the experiments the initially dry specimen was loaded up to a certain stress ratio and then wetted under the constant stress. For the dry state of the material the material parameters were calibrated based on the assumption of a constant solid hardness. The detailed calibration procedure applied is outlined in [60]. The values for the material parameters obtained are: $\varphi_c = 40°$, $h_{so} = 120$ [MPa], $n = 0.82$, $e_{io} = 0.3$, $e_{co} = 0.24$ and $\alpha = 0.18$. For the modeling of the time dependent process of degradation of the solid hardness caused by wetting the following additional material parameters were obtained:

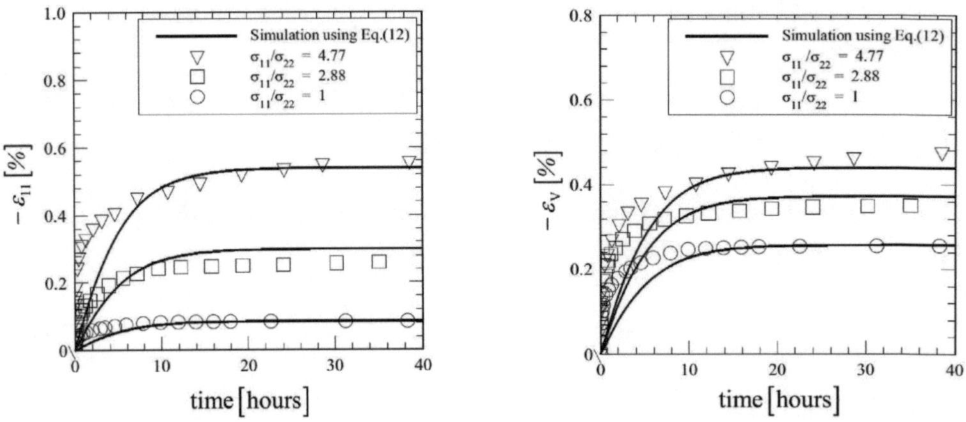

Figure 6.
Creep curves under a confining stress of $\sigma_{22} = -1200$ [kPa] and different stress ratios; the shapes indicate experimental data by Fu et al. [77]; solid curves are the numerical results obtained with respect to Eq.(13) for the degradation of the solid hardness.

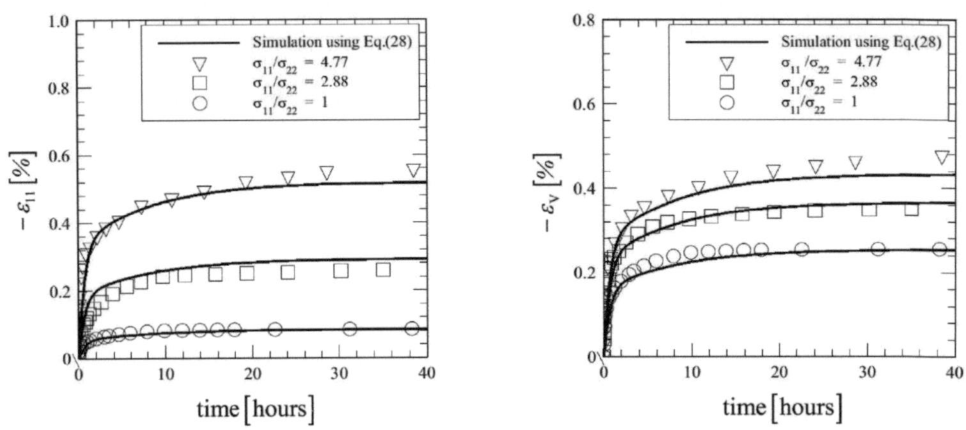

Figure 7.
Creep curves under a confining stress of $\sigma_{22} = -1200$ [kPa] and different stress ratios; the shapes indicate experimental data by Fu et al. [77]; solid curves are the numerical results obtained with respect to the extended Eq.(28) for the degradation of the solid hardness.

$$h_{sw} = 78.5 \text{ [MPa]}; \; c = 4.0 \text{ [h]}; \; \kappa = 0.6.$$

Numerical investigation with respect to the constitutive relation (13) show [60], that the prediction of the time dependent deformations immediately after wetting deviates from the experimental data. In particular, at initiation of creep the axial creep strain ε_{11} and the volume creep strain ε_V are underestimated as shown for a confining strass of $\sigma_{22} = -1200$ [kPa] and at three different stress ratios in **Figure 6**. It is obvious that the three material parameters h_{so}, h_{sw} and c of the constitutive relation (13) are not sufficient to capture both collapse settlements and long-time creep. For a refined modeling of wetting induced deformations the evolution Eq. (12) for the degradation of the solid hardness is extended and formulated as the sum of two parts [51], i.e.

$$\dot{h}_{st} = \dot{h}_{st\ collapse} + \dot{h}_{st\ creep}. \tag{28}$$

Herein $\dot{h}_{st\ collapse} = -\left(h_{st\ collapse} - h_{sw\ collapse}\right)/c_1$ denotes the part of the degradation of the solid hardness related to collapse deformation and

$\dot{h}_{st\ creep} = -(h_{st\ creep} - h_{sw\ creep})/c_2$ is the part related to long-term creep. As no clear distinction between collapse settlements and long-term creep can be detected from experiments, the constitutive parameters $c_1 < c_2$ and $h_{sw\ collapse} > h_{sw\ creep}$ can be calibrated using an optimization procedure. For consistency the difference of $h_{sw\ collapse}$ and $h_{sw\ creep}$ must be the same as the asymptotical value h_{sw} obtained with the constitutive Eq. (12), i.e. $h_{sw\ collapse} - h_{sw\ creep} = h_{sw}$. The quantities $1/c_1$ and $1/c_2$ control the velocities of the degradation of the solid hardnesses $h_{sw\ collapse}$ and $h_{sw\ creep}$, respectively. Thus, the quantity $1/c_1$ is related to the deformation velocity immediately after wetting, while the quantity $1/c_2$ is relevant of the long term creep. The values c_1 and c_2 can be optimized with respect to the experimental creep curve. The procedure leads for the evolution Eq. (28) the following material parameters:

$h_{sw\ collapse}$ = 90.5 [MPa]; c_1 = 0.6 [h]; $h_{sw\ creep}$ = 12.0 [MPa] and c_2 = 8.0 [h].

The results of numerical simulations using the extended evolution Eq. (28) are in good agreement with the experimental results as shown for the axial strain ε_{11} and volume strain ε_V in **Figure 7**.

5.2 Multistep creep

In order to simulation multistep creep initiated by repeated acceleration in the degradation of the solid hardness, e.g. caused by repeated rain water infiltration events into the rockfill body of the dam, two different concepts can be suggested for practical application. Either each event is modeled independent of the previous one, i.e. the values for parameters c and h_{sw} are calibrated for each individual creep curve, or subsequent events are modeled with respect of ongoing contributions of previous events. For the latter the rate of the solid hardness, $\dot{h}_{st\,(i)}$, of event (i) can be represented as:

$$\dot{h}_{st\,(i)} = \dot{h}_{st(i)collapse} + \sum_{k=1}^{i-1} \dot{h}_{st\,(k)}. \tag{29}$$

In Eq.(29) the terms on the right hand side consist of the contribution $\dot{h}_{st\,(i)collapse}$ initiated by a rapid change of the solid hardness in event (i), and the sum of contributions $\dot{h}_{st\,(k)}$ of previous events. Numerical simulations show that the accumulation concept proposed in Eq.(29) allows a more refined modeling of the whole history of creep and stress relaxation [51]. The application of the accumulation concept is illustrated in **Figure 8** based on the creep experiment by Sowers et al. [3]. To this end three sections can be distinguished. From the experimental data of section I the initially solid hardness h_{so} can be computed from the instantaneous settlement caused by the applied load of -0.8 MPa on the dry specimen. As no complete data for calibration are available the numerical simulations were carried out using an assumed value for the initial void ratio and estimated material parameters obtained by back analysis. In section II the creep deformation of the dry material can be modeled using the evolution Eq. (12) for the degradation of the solid hardness, $\dot{h}_{st\,\mathrm{II}}$. Herein the final value of the degraded solid hardness $h_{sw\,\mathrm{II}}$ can be obtained from Eq.(14) with respect to an estimated asymptotic value of the creep strain $\varepsilon_{V\infty\mathrm{II}}$. The degradation of the solid hardness in section III can be expressed as a combination of the influence of long-term creep in section II and the deformation immediately after flooding the specimen at time $t = 4500$minutes, i.e.

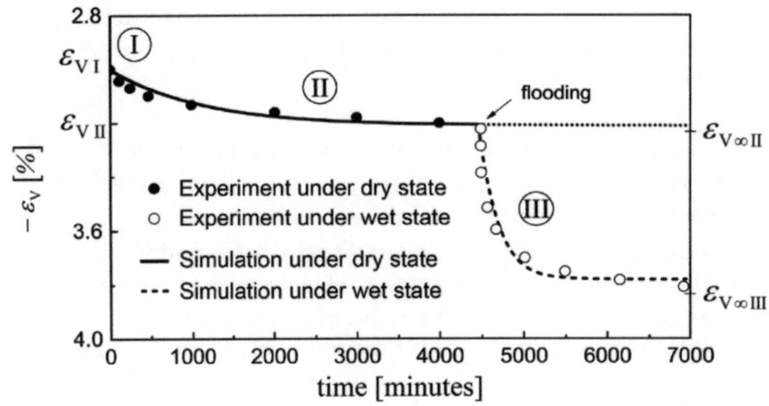

Figure 8.
Creep behavior of the stresses specimen under dry and wet conditions; experiments by Sowers et al. [3].

$\dot{h}_{st\,\mathrm{III}} = \dot{h}_{st\,\mathrm{III\,collapse}} + \dot{h}_{st\,\mathrm{II}}$. The value of $h_{sw\,\mathrm{III}}$ can be calculated from an estimation of the asymptotic value $\varepsilon_{V\infty\mathrm{III}}$. The creep velocities in part II and part III are different and the corresponding parameters c_{II} and c_{III} can be calibrated using Eq.(B5).

5.3 Influence of the stress deviator on the evolution of the creep deformation

The results of numerical simulations of the influence of the deviatoric stress on the stress strain relationship and on the creep behavior for a moisture sensitive broken sandstone are shown together with the experimental data in **Figure 9**. The 11 material parameter of the extended hypoplastic constitutive model were calibrated by

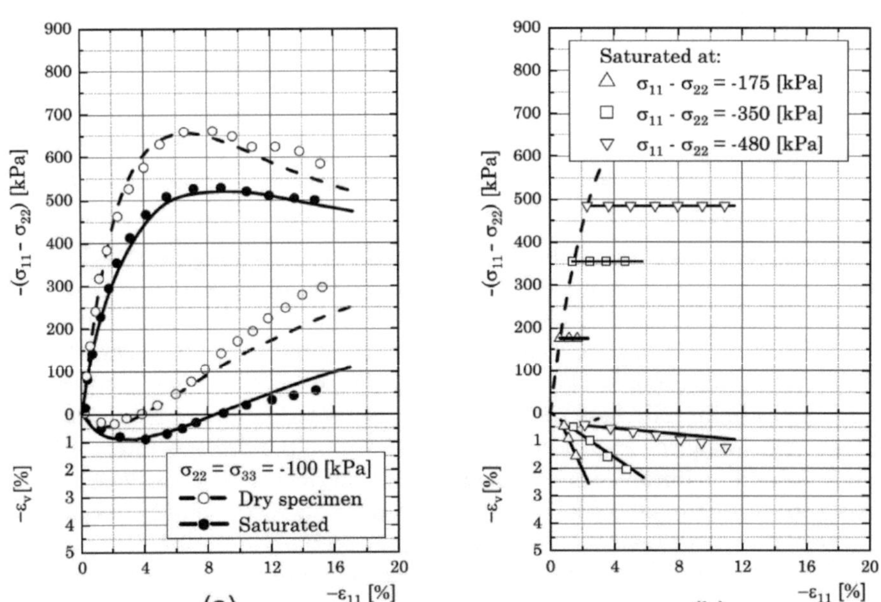

Figure 9.
Response of broken sandstone under triaxial compression: Dashed curves and solid curves are numerical responses for the dry and the water saturated specimens, respectively [50]; shapes denote the experimental data by Li [78].

Bauer et al. [50] using the experimental data obtained by Li [78]. As no creep behavior of the dry material in the triaxial compression test was reported, the calibration of the hypoplastic material parameters for the dry material were carried out for a constant solid hardness. In particular, the following set of material parameters were obtained for the initially dry material: $\varphi_c = 40°$, $h_{so} = 47[\text{MPa}]$, $n = 0.3$, $e_{io} = 0.59$, $e_{co} = 0.48$, $e_{do} = 0.20$, $\alpha = 0.18$ and $\beta = 2.50$. In the wet state the stressed material showed pronounced rheological behavior, which is taken into account in the numerical simulation by a degradation of the solid hardness using Eq.(13). For the saturated state of the material the additional material parameters for the extended hypoplastic model (25) are: $h_{sw} = 11.5[\text{MPa}]$, $\kappa = 0.7$ and $c = 3\text{days}$.

For monotonic triaxial compression under a lateral stress of $\sigma_{22} = -100$ [kPa] the numerical results are compared with the experimental data in **Figure 9a** for both dry and water saturated state of the material. As can be seen the course of the deviatoric stress $(\sigma_{11} - \sigma_{22})$ and the volumetric strain ε_V against the axial strain ε_{11} is different for the dry and the water saturated materials. The compaction at the beginning of deviatoric loading is higher and the subsequent dilatancy is much lower for the water saturated material. It can therefore be suspected that the critical void ratio will also be lower for the water saturated material. This is well captures with relation (20) for the critical void ratio depending on the mean stress and current state of the degraded solid hardness. The lower peak value of the deviatoric stress $(\sigma_{11} - \sigma_{22})$ and the subsequent strain softening is also well predicted with the same set of material parameters by the hypoplastic constitutive model proposed. In this context it can be noted that the peak value of $(\sigma_{11} - \sigma_{22})$ is not a material constant and is triggered in the extended constitutive Eq. (20) by the pressure dependent relative density factor f_d, which also depends on the current state of the degraded solid hardness h_{st}.

The behavior after water saturation at different deviatoric stresses is shown in **Figure 9b**. It is clearly visible that the axial creep strain ε_{11} is more pronounced for a higher level of the deviatoric stress. The corresponding volumetric creep paths, $\varepsilon_V(\varepsilon_{11})$, are almost linear and in good agreement with the experimental data. The inclination of $d\varepsilon_V/d\varepsilon_{11}$ decreases with an increase of the axial strain. The course of the stress ratio σ_{11}/σ_{22} against the inclination of the volumetric creep path, $d\varepsilon_V/d\varepsilon_{11}$,

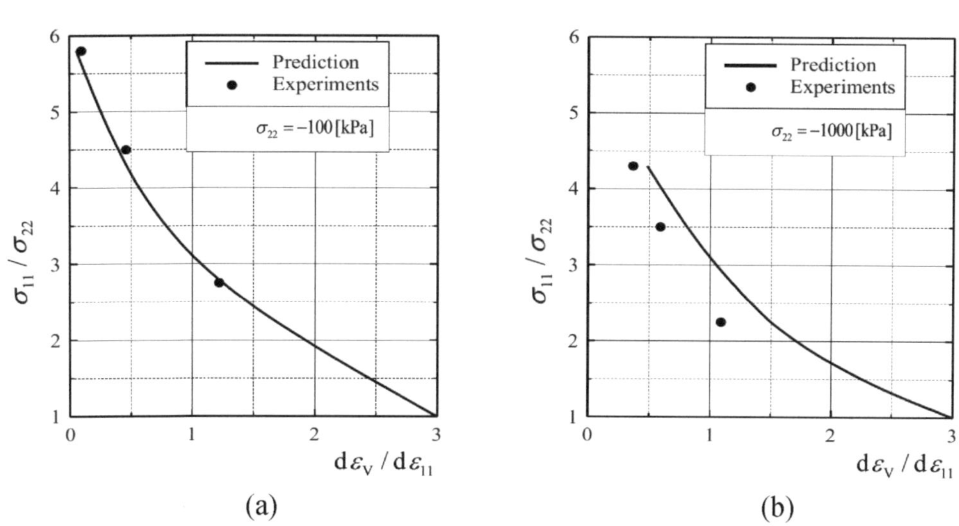

Figure 10.
Influence of the stress ratio σ_{11}/σ_{22} on the inclination of the volume strain curve $d\varepsilon_V/d\varepsilon_{11}$ under creep for a lateral stress of: (a) -100 [kPa]; and (b) -1000 [kPa]; the solid curves are numerical results; the shapes indicate experimental data by Li [78].

is nonlinear as shown in **Figure 10a**. With the same set of material parameters numerical simulations are also carried out for a ten times higher lateral stress, i.e. for $\sigma_{22} = -1000$ [kPa]. The corresponding relation between σ_{11}/σ_{22} and $d\varepsilon_V/d\varepsilon_{11}$ is shown in **Figure 10b**. Although the trend of the course is similar as for the lower lateral stress, there is a certain deviation between the numerical prediction and the experimental results. Numerical investigations show that the inclination of $d\varepsilon_V/d\varepsilon_{11}$ is mainly influenced by parameter κ of Eq.(25). The calibration of parameter κ based on the creep curve for a lower mean stress leads to predicted values of $d\varepsilon_V/d\varepsilon_{11}$, which are higher for a higher mean stress, i.e. the deviation to the experimental results increases with an increase of the mean stress. Fu et al. [32] showed that for different mean stresses and different stress ratios the experimental data can be largely approximated using a single hyperbolic function. The numerical simulations with the suggested approximation function, however, still show certain deviations from the experimental data. The investigations indicate that for a refined modeling parameter κ in Eq.(25) should be a function depending on the mean stress and the loading history. An appropriate modeling is still in progress.

6. Conclusions

In this paper the long-term behavior of weathered and moisture sensitive coarse-grained rockfill material is reviewed and the main mechanical properties are modeled in a phenomenological manner using a novel constitutive model based on the frame work of hypoplasticity. In order to take into account of the influence of grain crushing and the time dependent process of degradation of the strength of the solid material on the incremental stiffness, a so-called "solid hardness" is introduced as a state parameter into the constitutive model. The solid hardness is defined for monotonic isotropic compression of a grain assembly and it is a key parameter in the compression law by Bauer. It is shown that creep and stress relaxation are usually coupled. The calibration procedure of the material parameter relevant for the velocity of degradation of the solid hardness is outlined for different states at the creep curve and stress relaxation curve obtained in isotropic compression experiments. It is demonstrated that the strategy used for the embedding of the compression law into an extended 3-D hypoplastic continuum model can also be applied to other classes of constitutive models. With an enhanced hypoplastic constitutive model the influence of the state of weathering, the evolution of the degradation of the solid hardness, the packing density of the rockfill materials and the stress state on the incremental stiffness, the peak friction angle and the dilatancy angle can be modeled in a unified manner using a single set of material parameters. These properties of the model are confirmed by the comparison of numerical simulations with experimental data, in particular for triaxial compression tests under dry and wet conditions as well as for creep tests under different deviatoric stresses. It is shown that the evolution of the volume strain curve under creep strongly depends on the amount of the stress deviator where the rockfill material is saturated. The extended version of the evolution equation for the degradation of the solid hardness permits a refined modeling of collapse settlements and long-time creep caused by repeated changes of environmental conditions.

Acknowledgements

The author wishes to thank Dr. Z.Z. Fu, Dr. L. Li, Mr. M. Khosravi and Mr. S. Safikhani during their stay at the Institute of Applied Mechanics at Graz University of Technology for preparing parts of the numerical simulations shown in this paper.

Appendix A

As a heuristic example for the adaptation of the compression law (1) to a nonlinear elastic material description the following isotropic elastic material law is considered

$$\sigma_{ij} = K\,\varepsilon_{kk}\,\delta_{ij} + 2G\left(\varepsilon_{ij} - \frac{1}{3}\varepsilon_{ll}\delta_{ij}\right). \tag{A1}$$

Herein the bulk modulus, K, and shear modulus, G, are functions of invariants of chosen state variables such as the mean pressure or current void ratio. For isotropic compression Eq.(A1) leads for the mean pressure p:

$$p = -\frac{1}{3}\sigma_{ii} = -K\,\varepsilon_{kk}. \tag{A2}$$

With respect to the relationship $\mathrm{d}e = (1+e)\mathrm{d}\varepsilon_V$ one obtains for the volumetric strain $\varepsilon_V = \varepsilon_{kk}$ as a function of the current void ratio e and the initial void ratio $e_o = e(\varepsilon_V = 0;\ \ p = 0)$:

$$\varepsilon_V = \ln\left(\frac{1+e}{1+e_o}\right). \tag{A3}$$

Substitution of Eq.(A3) into Eq.(A2) leads to:

$$p = -K\,\ln\left(\frac{1+e}{1+e_o}\right) = K\ln\left(\frac{1+e_o}{1+e}\right). \tag{A4}$$

The compression law (1), i.e.

$$e = e_o\exp\left\{-\left(\frac{3p}{h_s}\right)^n\right\}$$

can be transformed to:

$$p = \frac{h_s}{3}\left[\ln\left(\frac{e_o}{e}\right)\right]^{(1/n)}. \tag{A5}$$

The comparison of relation (A4) with relation (A5) yields for the bulk modulus K depending on the current void ratio e:

$$K = \frac{h_s}{3}\,\frac{\left[\ln\left(\frac{e_o}{e}\right)\right]^{(1/n)}}{\ln\left(\frac{1+e_o}{1+e}\right)}. \tag{A6}$$

In the initial stress free state, i.e. $p = 0$, $\varepsilon_V = 0$ and $e = e_o$, the value of the bulk modulus is $K = 0$. With increasing compression the void ratio decreases, i.e. $e < e_o$, and the bulk modulus increases, $K > 0$, and for a vanishing void ratio, i.e. the volume cannot further compressed, the value of the bulk modulus tends asymptotically to $K \to \infty$ as $e \to 0$.

For the shear modulus, G, different functions are proposed in the literature. For instance, in the frequently applied empirical equation proposed by Hardin and

Drnevich [79] the shear modulus is a function of the current void ratio and the mean pressure, i.e.

$$G = G^* \frac{(2,97 - e)^2}{(1 + e)} \sqrt{\frac{p}{p_{atm}}}, \tag{A7}$$

where G^* is a material constant and p_{atm} denotes the atmospheric pressure.

Appendix B

The material parameter c can be calibrated either from the volume creep curve or from the stress relaxation curve. In the following the derivation of calibration procedures are demonstrated for four different cases. The quantities relevant for the calibration are illustrated in **Figure B1**. In **Case I** parameter c is related to the inclination of the volume creep curve at the beginning of creep, while in **Case II** the inclination of the volume creep curve at $\varepsilon_{V\infty}/2$ is considered. **Case III** and **Case IV** deals with the calibration of parameter c at two different states of the stress relaxation curves

B.1 Case I

Substituting the evolution Eq. (12) for the solid hardness into Eq.(5) one obtains for the rate of the volume strain under creep:

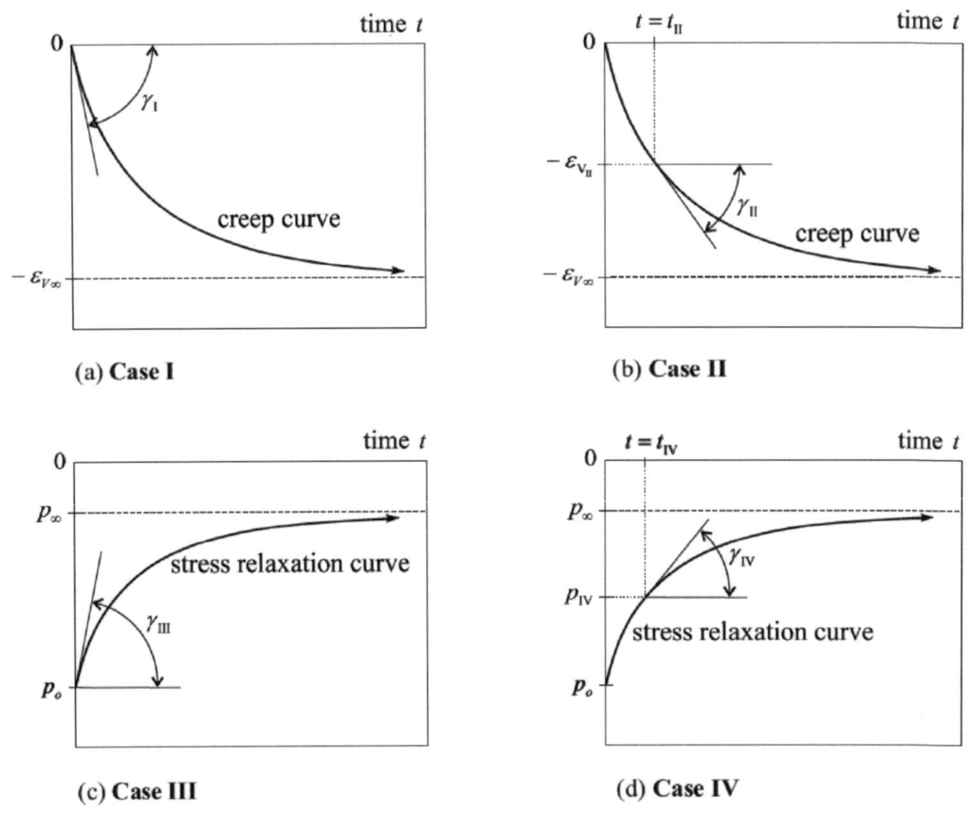

Figure B1.
Quantities at creep and stress relaxation curves relevant for the calibration.

$$\frac{\mathrm{d}\varepsilon_V}{\mathrm{d}t} = \dot{\varepsilon}_V = -\left(\frac{1}{c}\right)\frac{n\,e}{1+e}\left(\frac{3p_o}{h_{st}}\right)^n\left(1-\frac{h_{sw}}{h_{st}}\right). \tag{B1}$$

Herein the degraded solid hardness h_{sw} is defined for $h_{st}(t \to \infty) = h_{sw}$.

In **Case I** the inclination γ_{I} of the volume creep curve at the reference time $t = 0$ depends on the corresponding values of state quantities $e(t = 0) = e_{\mathrm{I}}$; $p(t = 0) = p_o$ and $h_{st}(t = 0) = h_{st\,\mathrm{I}}$. From:

$$\tan\gamma_{\mathrm{I}} = -\frac{\mathrm{d}\varepsilon_V}{\mathrm{d}t}\bigg|_{t=0} = \left(\frac{1}{c}\right)\frac{n\,e_{\mathrm{I}}}{(1+e_{\mathrm{I}})}\left(\frac{3p_o}{h_{st\,\mathrm{I}}}\right)^n\left(1-\frac{h_{sw}}{h_{st\,\mathrm{I}}}\right), \tag{B2}$$

one obtains for parameter c:

$$c = \frac{n\,e_{\mathrm{I}}}{\tan\gamma_{\mathrm{I}}(1+e_{\mathrm{I}})}\left(\frac{3p}{h_{st\,\mathrm{I}}}\right)^n\left(1-\frac{h_{sw}}{h_{st\,\mathrm{I}}}\right). \tag{B3}$$

B.2 Case II

In **Case II** the inclination γ_{II} of the volume creep curve is considered at time $t = t_{\mathrm{II}}$ where $\varepsilon_{V_{\mathrm{II}}} = \varepsilon_{V\infty}/2$; $e_{\mathrm{II}} = e(t = t_{\mathrm{II}})$ and $h_{st\,\mathrm{II}} = h_{st}(t = t_{\mathrm{II}})$. The asymptotic value $\varepsilon_{V\infty}$ can be estimated from the experiment. Then the corresponding relation for parameter c can be obtained from:

$$\tan\gamma_{\mathrm{II}} = -\frac{\mathrm{d}\varepsilon_V}{\mathrm{d}t}\bigg|_{t=t_{\mathrm{II}}} = \left(\frac{1}{c}\right)\frac{n\,e_{\mathrm{II}}}{(1+e_{\mathrm{II}})}\left(\frac{3p_o}{h_{st\,\mathrm{II}}}\right)^n\left(1-\frac{h_{sw}}{h_{st\,\mathrm{II}}}\right) \tag{B4}$$

to:

$$c = \frac{n\,e_{\mathrm{II}}}{\tan\gamma_{\mathrm{II}}(1+e_{\mathrm{II}})}\left(\frac{3p}{h_{st\,\mathrm{II}}}\right)^n\left(1-\frac{h_{sw}}{h_{st\,\mathrm{II}}}\right). \tag{B5}$$

B.3 Case III

For stress relaxation under constant volume staring from $p(t = 0) = p_o$ the rate of the mean stress according to Eq.(10) reads:

$$\dot{p} = p_o\left(\frac{\dot{h}_{st}}{h_{st}}\right). \tag{B6}$$

Substituting the evolution relation (12) for the solid hardness into Eq.(B6) one obtains:

$$\frac{\mathrm{d}p}{\mathrm{d}t} = \dot{p} = -\frac{p_o}{c}\left(1-\frac{h_{sw}}{h_{st}}\right). \tag{B7}$$

The inclination γ_{III} of the stress relaxation curve at the reference time $t = 0$ depends on the quantities $p(t = 0) = p_o$ and $h_{st}(t = 0) = h_{so}$. From:

$$\tan\gamma_{\mathrm{III}} = -\frac{\mathrm{d}p}{\mathrm{d}t}\bigg|_{t=0} = \frac{p_o}{c}\left(1-\frac{h_{sw}}{h_{so}}\right) \tag{B8}$$

and with respect of Eq.(15), i.e. $h_{sw} = h_{so} p_\infty / p_o$, one obtains for parameter c:

$$c = \frac{p_o}{\tan \gamma_{III}} \left(1 - \frac{p_\infty}{p_o}\right), \tag{B9}$$

where p_∞ denotes the asymptotic value of the stress relaxation curve for $t \to \infty$.

B.4 Case IV

As in experiments the inclination γ_{III} of the stress relaxation curve at the reference time $t = 0$ is difficult to measure an alternative calculation of parameter c is outlined in the following at $t = t_{IV}$ and $p_{IV} = (p_o + p_\infty)/2$. Then the corresponding rate of the solid hardness reads:

$$\dot{p}_{IV} = -\frac{p_{IV}}{c} \left(1 - \frac{h_{sw}}{h_{st\,IV}}\right), \tag{B10}$$

with:

$$h_{st\,IV} = h_{st}|_{t=t_{IV}} = h_{so} \frac{p_{IV}}{p_o} = \frac{h_{so}}{2} \left(1 + \frac{p_\infty}{p_o}\right). \tag{B11}$$

Parameter c can be obtained from:

$$\tan \gamma_{IV} = -\frac{dp}{dt}\bigg|_{t=t_{IV}} = \frac{p_{IV}}{c} \left(1 - \frac{h_{sw}}{h_{so}} \frac{p_o}{p_{IV}}\right) \tag{B12}$$

to:

$$c = \frac{(p_o + p_\infty)}{2 \tan \gamma_{IV}} \left(1 - \frac{2 p_\infty}{(p_o + p_\infty)}\right). \tag{B13}$$

Appendix C

In order to embed the compression law (1) in a consistent manner into the hypoplastic constitutive Eq. (16) the rate of the mean pressure, i.e. $\dot{p} = -\dot{\sigma}_{ii}/3$, under monotonic isotropic compression starting from the maximum void ratio e_{i0} must coincide for both equations. For this particular compression path the void ratio $e(p) = e_i(p)$ and the scalar factors f_d, f_e, f_σ and $\hat{a}$ of the constitutive Eq. (16) result in the following values

$$f_d(e = e_i) = \left(\frac{e_{i0} - e_{d0}}{e_{c0} - e_{d0}}\right)^\alpha, \; f_e(e = e_i) = 1, \; f_\sigma(\sigma_{11} = \sigma_{22} = \sigma_{33}) = 3,$$
$$\text{and } \hat{a}(\sigma_{11} = \sigma_{22} = \sigma_{33}) = \frac{\sqrt{8/3}\, \sin \varphi_c}{3 - \sin \varphi_c}. \tag{C1}$$

With respect to the relations in (C1) the rate of the mean pressure calculated from the hypoplastic constitutive Eq. (16) reads:

$$\dot{p}_{hypo} = -f_b \frac{\dot{e}_i}{3(1+e_i)} \left[\frac{8\sin^2\varphi}{(3-\sin\varphi)^2} + 1 - \frac{2\sqrt{2}\sin\varphi}{(3-\sin\varphi)}\left(\frac{e_{i0}-e_{d0}}{e_{c0}-e_{d0}}\right)^{\alpha}\right]. \quad \text{(C2)}$$

From the compression law (1) one obtains:

$$\dot{p}_{c\_law} = -\frac{\dot{e}_i}{e_i}\frac{h_s}{3n}\left(\frac{3p}{h_s}\right)^{(1-n)}. \quad \text{(C3)}$$

From the consistency condition $\dot{p}_{hypo} = \dot{p}_{c\_law}$, factor f_b can be determined, i.e.:

$$f_b = \frac{h_s\,(1+e_i)}{n\,e_i}\left[\frac{8\sin^2\varphi}{(3-\sin\varphi)^2} + 1 - \frac{2\sqrt{2}\sin\varphi}{(3-\sin\varphi)}\left(\frac{e_{i0}-e_{d0}}{e_{c0}-e_{d0}}\right)^{\alpha}\right]^{-1}\left(\frac{3p}{h_s}\right)^{(1-n)}. \quad \text{(C4)}$$

Appendix D – Symbols

D.1 State quantities

σ_{ij}	Effective Cauchy stress
e	Current void ratio
h_{st}	Solid hardness depending on the current state of weathering

D.2 Constitutive quantities

$\overset{\nabla}{\sigma}_{ij}$	Objective stress rate
$\dot{\varepsilon}_{ij}$	Strain rate
$\dot{\varepsilon}_V$	Volume strain rate, i.e. $\dot{\varepsilon}_V = \dot{\varepsilon}_{kk} = \dot{\varepsilon}_{11} + \dot{\varepsilon}_{22} + \dot{\varepsilon}_{33}$
$\dot{h}_{st}$	Degradation rate of the solid hardness
p	Mean effective pressure, i.e. $p = -\sigma_{kk}/3 = -(\sigma_{11} + \sigma_{22} + \sigma_{33})/3$
$\dot{p}$	Rate of the mean effective pressure
$\dot{e}$	Rate of the void ratio
e_i	Pressure dependent maximum void ratio
e_c	Pressure dependent critical void ratio
e_d	Pressure dependent minimum void ratio

D.3 Material parameters

h_s	Solid hardness of an unweathered material
h_{sw}	Asymptotical value of the degraded solid hardness for $t \to \infty$
c	Velocity dependent material parameter in the degradation law of the solid hardness
e_{i0}	Maximum void ratio at zero stress
e_{c0}	Critical void ratio at zero stress
e_{d0}	Minimum void ratio at zero stress
n	Material parameter in the compression law by Bauer
α	Material parameter of the hypoplastic constitutive equation
β	Material parameter of the hypoplastic constitutive equation

φ_c Critical friction angle
κ Material parameter of the enhanced hypoplastic constitutive equation

D.4 Initial values

e_o Initial void ratio
p_o Effective reference pressure
h_{so} Initial solid hardness at the reference time $t = 0$

Author details

Erich Bauer
Institute of Applied Mechanics, Graz University of Technology, Graz, Austria

*Address all correspondence to: erich.bauer@tugraz.at

References

[1] Justo J.L., Durand P. Settlement-time behaviour of granular embankments. International Journal for Numerical and Analytical Methods in Geomechanics. 2000, 24(3): 281–303.

[2] Hunter G., Fell R. Rockfill modulus and settlement of concrete face rockfill dams. Journal of Geotechnical and Geoenvironmental Engineering. 2003, 129(10): 909–917.

[3] Sowers G. F., Williams R. C., Wallace T. S. Compressibility of broken rock and settlement of rockfills. Proc. 6th ICSMFE, Montreal. 1965, Vol. 2: 561–565.

[4] Nobari E. S., Duncan J. M. Effect of reservoir filling on stresses and movements in earth and rockfill dams. Department of Civil Engineering, Report No. TE-72-1. University of California. 1972.

[5] Marsal R.J. Mechanical properties of rockfill. In: Embankment Dam Engineering. (eds.) Casagrande. R.C. Hirschfeld and S.J. Poulos, John Wiley & Sons, New York. 1973: 109–200.

[6] Alonso E., Oldecop L.A. Fundamentals of rockfill collapse. Proc. of the 1st Asian Conf. on Unsaturated Soils, eds. Rahardjo H., Toll DG., Leong EC., Balkema Press, Rotterdam. 2000, 3–13.

[7] Alonso E. E., Cardoso R. Behaviour of Materials for Earth and Rockfill Dams: Perspective from Unsaturated Soil Mechanics. Proceedings of the 2nd Int. Conf. on Long Term Behaviour of Dams, (eds.) Bauer, Semprich and Zenz, Publisher by Verlag der Technischen Universität Graz, ISBN 978–3–85125-070-1, 2009: 1–38.

[8] Wei Z., Xiaolin C., Chuangbing Z., Xinghong L. Creep analysis of high concrete-faced rockfill dam. International Journal for Numerical Methods in Biomedical Engineering. 2010, 26(11): 1477–1492.

[9] Brauns J., Kast K., Blinde A. 1980. Compaction Effects on the Mechanical and Saturation Behaviour of Disintegrated Rockfill. Proceedings of International Conference on Compaction, Paris. 1980, Vol. 1: 107–112.

[10] Oldecop L. A., Alonso E.E. 2007. Theoretical Investigation of the Time-Dependent Behaviour of Rockfill. Géotechnique. 2007, Vol. 57, N.3: 289–301.

[11] Oldecop L. A., de Agreda E. Testing Rockfill under Relative Humidity Control. Geotechnical Testing Journal. 2004, Vol. 27, N.3: 269–278.

[12] Zhang B. Y., Wang J. G., Shi R. F. Time-dependent deformation in high concrete-faced rockfill dam and separation between concrete face slab and cushion layer. Comput. Geotech. 2004, 31(7): 559–573.

[13] Modares M., Quiroz J.E. Structural analysis framework for concrete-faced rockfill dams. International Journal of Geomechanics. 2016, 16(1): 1–14.

[14] Cen W.J., Wen L.S., Zhang Z.Q., Xiong K. Numerical simulation of seismic damage and cracking of concrete slabs of high concrete face rockfill dams. Water Science and Engineering. 2016, 9(3): 205–211.

[15] Khosravi M., Li L., Bauer E. Numerical simulation of post-construction deformation of a concrete face rockfill dam. Proceedings of the 4th International Conference on Long-Term Behaviour and Environmentally Friendly Rehabilitation Technologies of Dams, LTBD 2017, (eds.) Noorzad, A.,

Bauer, E., Ghaemian, M., Ebrahimian, B., Published by Verlag der Technischen Universität Graz, ISBN 978–3-85125-564-5, 2017: 307–314.

[16] Saberi M, Annan CD, Konrad JM. Numerical analysis of concrete-faced rockfill dams considering effect of face slab – cushion layer interaction. Canadian Geotechnical Journal. 2018, 55 (10): 1489–1501.

[17] Marsal R.J. Large-scale testing of rockfill materials. Journal of the Soil Mechanics and Foundations Division, ASCE. 1967, 93(2): 27–43.

[18] Hu W., Frossart E., Hicher P. Y., Dano C. A new method to evaluate the mechanical behavior of granular material with large particles. Theory and validation. Proceedings of the 2nd Int. Conf. on Long Term Behaviour of Dams, (eds.) Bauer, Semprich and Zenz, Publisher by Verlag der Technischen Universität Graz, ISBN 978–3–85125-070-1, 2009: 437–478.

[19] Ovalle C., Frossard E., Dano C., Hu W., Maiolino S., Hicher P.Y. The effect of size on the strength of coarse rock aggregates and large rockfill samples through experimental data. Acta Mechanica. 2014, Vol. 225(8): 2199–2216.

[20] Ham T.G., Nakata Y., Orense R., ASCE, M., Hyodo M. Influence of Water on the Compressive Behavior of Decomposed Granite Soil. Journal of Geotechnical and Geoenvironmental Engineering. 2010, Vol. 136, (5): 697–705.

[21] Zhang B., Chen T., Peng C., Qian X., Jie Y. Experimental Study on Loading-Creep Coupling Effect in Rockfill Material. Int. J. Geomech. 2017, 17(9): 04017059. DOI: 10.1061/(ASCE)GM.1943-5622.0000938

[22] Zhou X., Chi S., Jia Y. Wetting deformation of core-wall rockfill dams. Int. J. Geomechanics. 2019, 19(8), 4019084. https://doi.org/10.1061/(ASCE)GM.1943-5622.0001444. https://doi.org/10.1061/(ASCE)GM.1943-5622.0001444

[23] Yin Y., Wu Y., Zhang B., Ding Y., Sun X. Two-stage wetting deformation behaviour of rock-fill material. Environmental Geotechnics. 2019, https://doi.org/10.1680/jenge.18.00130

[24] Viswanath P., Das A. Behavioural Study on Geomaterial Undergoing Chemo-Mechanical Degradation. In: Prashant A., Sachan A., Desai C. (eds). Advances in Computer Methods and Geomechanics. Lecture Notes in Civil Engineering. 2020, Vol 55. Springer, Singapore. 305–314. https://doi.org/10.1007/978-981-15-0886-8_25

[25] Terzaghi K. Discussion on salt springs and lower bear river dams, Trans. ASCE 125 (2). 1960: 139–148.

[26] Ovalle C., Dano C., Hicher P.Y., Cisternas M. Experimental framework for evaluating the mechanical behavior of dry and wet crushable granular materials based on the particle breakage ratio. Canadian Geotechnical Journal. 2015, Vol. 52: 587–598, https://doi.org/10.1139/cgj-2014-0079.

[27] Brauns J., Kast K., Blinde A. Compaction Effects on the Mechanical and Saturation Behavior of Disintegrated Rockfill. Proc. Int. Conf. on Compaction, Paris. 1980, Vol. 1: 107–112.

[28] Kast K. Mechanisches Verhalten von Granitschuettungen. Veröffentlichungen des Institutes für Bodenmechanik und Felsmechanik der Universität Fridericiana in Karlsruhe, Heft 125. 1992. (in German)

[29] Oldecop L.A., Alonso E.E. A model for rockfill compressibility. Géotechnique. 2001, Vol. 51, N.2: 127–139.

[30] Zhou W., Hua J., Chang X.L., Zhou C. Settlement analysis of the Shuibuya concrete-face rockfill dam. Computers and Geotechnics. 2011, 38: 269–280.

[31] Hu W., Yin Z., Dano C., Hicher P.Y. A constitutive model for granular materials considering grain breakage. Science China Technological Sciences. 2011, 54(8): 2188–2196.

[32] Fu Z.Z., Chen S.S., Liu S.H. Hypoplastic constitutive modelling of the wetting induced creep of rockfill materials. Sci China Tech Sci. 2012, 55: 2066–2082.

[33] Frossard E., Dano C., Hu W. Rockfill shear strength evaluation: A rational method based on size effects. Géotechnique. 2012, 62(5): 415–427.

[34] Xiao Y., Liu H., Chen Y. Testing and modeling of the state-dependent behaviors of rockfill material. Computers and Geotechnics. 2014, 61: 153–165.

[35] Xiao Y., Liu H., Chen Y., Jiang J. Bounding Surface Plasticity Model Incorporating the State Pressure Index for Rockfill Materials. J. Eng. Mech. 2014, 140(11): 04014087. DOI:10.1061/(ASCE)EM.1943-7889.0000802

[36] Alonso E. E. Fracture mechanics and rockfill dams". Soils and Rocks, An International Journal of Geotechnical and Geoenvironmental Engineering, São Paulo. 2014, 37(1): 3–35.

[37] Fu Z., Chen S., Peng C. Modeling cyclic behavior of rockfill materials in a framework of generalized plasticity. International Journal of Geomechanics. 2014, DOI:10.1061/(ASCE)GM.1943-5622.0000302, 191–204

[38] Wen L., Chai J., Xu Z., Qin Y., Li Y. Monitoring and numerical analysis of behaviour of Miaojiaba concrete-face rockfill dam built on river gravel foundation in China. Computers and Geotechnics. 2017, 85: 230–248.

[39] Tosun H., Tosun T.V. Dynamic Analysis of Embankment Dams under Strong Seismic Excitation and a Case Study. Proceedings of the 4th International Conference on Long-Term Behaviour and Environmentally Friendly Rehabilitation Technologies of Dams, LTBD 2017. Noorzad, A., Bauer, E., Ghaemian, M., Ebrahimian, B. (eds.), Published by Verlag der Technischen Universität Graz, ISBN 978–3–85125-564-5 2017: 747–753.

[40] Xu M., Jin D., Song E., Shen D. A rheological model to simulate the shear creep behavior of rockfills considering the influence of stress states Acta Geotechnica. 2018, 13:1313–1327. https://doi.org/10.1007/s11440-018-0716-8

[41] Wen L., Chai J., Xu Z., Qin Y.,Li Y. A statistical review of the behaviour of concrete-face rockfill dams based on case histories. Geotechnique. 2018, 68 (9): 749–771.

[42] Fu Z.Z., Chen S.S., Wie K.M. A generalized plasticity model for the stress-strain and creep behavior of rockfill materials. Science China Technological Sciences. 2019, 62, 649–664. Doi: 10.1007/s11431-018-9362-3

[43] Xu M., Jin D., Song E., Shen Z., Yang Z., Fu J. Full-scale creep test and back-analysis of the long-term settlement of heavy-loaded shallow foundations on a high rockfill embankment. Computers and Geotechnics. 2019, 115: 103156.

[44] Yao F.H., Guan S.H., Yang H., Chen Y., Qiu H.F., Ma G., Liu Q.W. Long-term deformation analysis of Shuibuya concrete face rockfill dam based on response surface method and improved genetic algorithm. Water Science and Engineering. 2019, 12(3): 196–204.

[45] Liu S., Sun Y., Shen C., Yin Z. Practical nonlinear constitutive model for rockfill materials with application to rockfill dam. Computers and Geotechnics. 2020, https://doi.org/10.1016/j.compgeo.2019.103383.

[46] Zhou X., Chi S., Wang M., Jia Y. Study on wetting deformation characteristics of coarse granular materials and its simulation in core-wall rockfill dams. Int J Numer Anal Methods Geomech. 2020, 44: 851–873. DOI: 10.1002/nag.3042

[47] Bauer E. Hypoplastic Modelling of Moisture-Sensitive Weathered Rockfill Materials. Acta Geotechnica. 2009, Vol. 4, (4): 261–272.

[48] Bauer E., Fu Z., Liu S. Constitutive Modelling of Rheological Properties of Materials for Rockfill Dams. In: Proceedings of the 6th International Conference on Dam Engineering, Lisbon, Portugal, February 15–17, 2011, eds. C. Pina, E. Portela, J. Gomes. Published by: CI-Premier Pte Ltd (Singapore), ISBN: 978–981–08-7896-2 (CD ROM): 1–14.

[49] Bauer E., Fu Z. Modeling of weathered and moisture sensitive granular materials. In: Advances in Bifurcation and Degradation in Geomaterials, Proceedings of the 9th International Workshop on Bifurcation and Degradation in Geomaterials, eds. S. Bonelli, C. Dascalu, F. Nicot. Springer Series in Geomechanics & Geoengineering, ISBN 978–94–007-1420-5, e-ISBN: 978–94–007-1421-2. 2011: 331–336.

[50] Bauer E., Fu Z., Liu S. Influence of pressure and density on the rheological properties of rockfills. Frontiers of Structural and Civil Engineering, 6, DOI 10.1007/s11709-012-0143-0. 2012: 25–34.

[51] Bauer E. Constitutive Modelling of Wetting Deformation of Rockfill Materials. International Journal of Civil Engineering. 2019, 17: 481–486. https://doi.org/10.1007/s40999-018-0327-7, Online ISSN 2383–3874, Print ISSN 1735–0522

[52] Bauer E., Safikhani S., Li L. Numerical simulation of the effect of grain fragmentation on the evolution of microstructure quantities. Meccanica. 2019, 54(4): 631–642. DOI 10.1007/s11012-019-00953-0 , ISSN 0025–6455

[53] Fu Z.Z., Bauer E. Hypoplastic Constitutive Modelling of the Long Term Behaviour and Wetting Deformation of Weathered Granular Materials. Proc. of the 2nd Int. Conference on Long Term Behaviour of Dams, Graz, Austria, eds. Bauer, E., Semprich, S. Zenz, G., ISBN: 978–3–85125-070-1. 2009: 473–478.

[54] Bauer E., Fu Z., Liu S. Hypoplastic Constitutive Modeling of Wetting Deformation of Weathered Rockfill Materials. Frontiers of Architecture and Civil Engineering in China. 2010, Vol. 4, (1): 78–91.

[55] Fu Z.Z., Chen S.S., Liu S.H. Hypoplastic constitutive modelling of the wetting induced creep of rockfill materials. Science China Technological Sciences. 2012, 55, (7): 2066–2082.

[56] Khosravi M., Li L., Safikhani S., Bauer E. Numerical simulations of the wetting effect on the long-term behavior of concrete face rockfill dams. Proceedings of the 5th International Symposium on Dam Safety, Istanbul, Turkey. 2018: 989–997.

[57] Bauer E. Constitutive Modelling of Critical States in Hypoplasticity. In: Proceedings of the Fifth International Symposium on Numerical Models in Geomechanics, Davos, Switzerland, Balkema. 1995: 15–20.

[58] Bauer E. Calibration of a comprehensive hypoplastic model for

granular materials. Soils and Foundations. 1996, 36(1): 13–26.

[59] Bauer E., Huang W., Wu W. Investigation of shear banding in an anisotropic hypoplastic material. Solids and Structures. 2004, 41: 5903–5919.

[60] Li L., Wang Z., Liu S., Bauer E. Calibration and performance of two different constitutive models for rockfill materials. Water Science and Engineering. 2016, 9(3): 227–239. http://dx.doi.org/10.1016/j.wse.2016.11.005

[61] Tejchman J., Bauer E. Numerical simulation of shear band formation with a polar hypoplastic constitutive model. Comput. Geotech. 1996, 19: 221–44.

[62] Huang W., Nübel K., Bauer E. A polar extension of hypoplastic model for granular material with shear localization. Mechanics of Materials. 2002, 34: 563–576.

[63] Huang W., Bauer E. Numerical investigations of shear localization in a micro-polar hypoplastic material. Int. J. for Numerical and Analytical Methods in Geomechanics. 2003, 27: 325–352.

[64] Ebrahimian B., Bauer E. Numerical simulation of the effect of interface friction of a bounding structure on shear deformation in a granular soil. Int. J. Numer. Anal. Meth. Geomech. 2012, 36 (12): 1486–1506.

[65] Bauer E., Li L., Huang W. Hypoplastic Constitutive modelling of grain damage under plane shearing. In Bifurcation and Degradation of Geomaterials in the New Millennium. 2015: 181–187.

[66] Bauer E., Li L., Khosravi M. Modelling grain damage under plane strain compression using a micro-polar continuum". In Proceedings of the 11th International Workshop on Bifurcation and Degradation in Geomaterials Dedicated to Hans Muhlhaus, Limassol, Cyprus, 21–25 May 2017, Papamichos, E., Papanastasiou, P., Pasternak, E., Dyskin, A. (eds.), Springer Series in Geomechanics and Geoengineering, DOI 10.1007/978-3-319-56397-8. 2017: 539–546.

[67] Bauer E. Modelling Grain Fragmentation in Hypoplasticity. Springer Nature Switzerland AG 2019, W. Wu (Ed.): Desiderata Geotechnica, SSGG. 2019: 1–20.

[68] Bauer E., Safikhani S. Numerical Investigation of Grain Fragmentation of a Granular Specimen under Plane Strain Compression. ASCE, Int. J. Geomech. 2020, 20(3): 04020007 DOI: 10.1061/(ASCE)GM.1943-5622.0001608.

[69] Gudehus G. A Comprehensive Constitutive Equation for Granular Materials. Soils and Foundations. 1996: 36, (1): 1–12.

[70] Yamamuro J.A., Bopp P.A., Lade P. V. One-dimensional compression of sands at high pressures. J. Geotech. Engrg., ASCE. 1996, 122 (2): 147–154.

[71] Laufer I. Grain crushing and high-pressure oedometer tests simulated with the discrete element method. Granul. Matter. 2015, 17(3): 389–412. http:// dx.doi.org/10.1007/s10035-015-0559-z.

[72] Fu R., Hua X., Zhou B. Discrete element modeling of crushable sands considering realistic particle shape effect. Computers and Geotechnics. 2017, 91, 179–191.

[73] Jaky J. Pressure in silos. Proc. 2nd International Conference on Soil Mechanics and Foundation Engineering, Rotterdam, Nederland. 1948, Vol. 1: 103–107.

[74] Matsuoka H., Nakai T. Stress-Strain Relationship of Soil Based on the SMP. Proceedings of 9th ICSMFE, Constitutive Equations of Soils. 1977: 153–162.

[75] Schofield A.N., Wroth C.P. Critical State Soil Mechanics, McGraw-Hill, London. 1968.

[76] Bauer E. Conditions for embedding Casagrande's critical states into hypoplasticity. Mechanics of Cohesive-Frictional Materials. 2000 Vol. 5: 125–148.

[77] Fu H., Ling H. Experimental research on the engineering properties of the fill materials used in the Cihaxia concrete faced rockfill dam. Research Report, Nanjing Hydraulic Research Institute, Nanjing, China. 2009 (in Chinese).

[78] Li G.X. Triaxial wetting experiments on rockfill materials used in Xiaolangdi earth dam. Research report from Tsinghua University. 1988.

[79] Hardin B. O., Drnevich V. P. Shear Modulus and Damping in Soils: Design Equations and Curves. J. Soil Mech. Found. Div. Proc. ASCE. 1972, Vol. 98, No. 7: 667–692.

Chapter 5

A State-Dependent Constitutive Model for Unsaturated Rockfill Materials

Liujiang Wang and Zhongzhi Fu

Abstract

This chapter presents a state-dependent elastoplastic constitutive model for both saturated and unsaturated rockfill materials. The model, which is developed within an extended critical-state framework, uses two independent stress state variables: total stress and total suction. The loading-collapse (LC) curve proposed by Oldecop and Alonso for unsaturated rockfills is used herein. A unified hardening parameter, which could consider the effects of stress level, internal state (density) and relative humidity, is introduced to describe the state-dependent dilatancy of saturated and unsaturated rockfill materials. The details of the model formulation and parameters determination are described and reported. Numerical simulations on the triaxial tests, such as the drained shear tests on the saturated specimens with different initial dry densities, shear tests on the specimens with different relative humidity and wetting deformation tests under constant vertical strain rate, have been carried out using the proposed model. The numerical results show that the stress-strain relationships at both loose and dense, saturated and unsaturated states can be properly modelled with a single set of parameters. Additionally, the proposed model can also capture some other key features such as the strain-softening behaviour at the dense state and low confining stress, the sudden stress relaxing subjected to the flooding under a constant vertical strain.

Keywords: constitutive relations, unsaturated rockfills, collapse deformation, initial density

1. Introduction

It is well recognised from the experimental studies and engineering practice that the influence of water on the mechanical behaviour of rockfill materials is significant. A lot of laboratory experiments carried out in the past showed that the rockfill materials may undergo significant amounts of strain upon flooding [1–10]. This phenomenon is usually called 'collapse deformation', referring to a strain increment not related to the load increment. On the other hand, a remarkable influence of the compacted density and confining pressure on the strength and deformation behaviours of rockfill materials was observed in some large-scale triaxial tests [11–15]. The experimental results on rockfill materials at a dense state showed the remarkable strain-softening and dilatancy behaviours, while the rockfill materials at a loose state exhibited the strain hardening and volumetric contraction behaviours. This state-dependent behaviour was also observed in the triaxial tests on rockfill materials used in Guanyinyan dam that will be introduced in the next section. Furthermore, compared with the fully saturated rockfill materials, a more obvious state-dependent behaviour was observed for unsaturated rockfills in the relative humidity-controlled triaxial tests [7, 16]. The rockfills subjected to the larger suction exhibited more remarkable strain-softening and dilatancy at a low confining pressure or with a small initial void ratio. Thus, presenting a constitutive model which could consider the collapse deformation and state-dependent behaviour for unsaturated rockfill materials is necessary.

Rockfill materials have been widely used in the constructions of rockfill dams, and they should qualitatively be defined very well before beginning to the design stage [17–20]. The collapse settlement of upstream shell for clay core rockfill dam occurs during the impounding, which has attracted extensive attentions from both engineers and scientists since 1970s [3], for it may cause cracks on the dam crest and result in damage to impermeable systems. Thereby, proposing a satisfactory technique for the prediction of collapse settlement is clearly important. Nobari and Duncan [3] first introduced collapse effects into finite element (FE) analysis to predict collapse settlement in rockfill dams. This calculation starts by performing the FE analysis using a set of material parameters corresponding to the dry conditions; then the collapse is numerically simulated in two stages. First, the stress change caused by the saturation at the constant strain is determined using the experimental data. In the second stage, the nodal forces that restore the equilibrium are applied, and the computed strains reproduce the collapse strains. Naylor et al. [21] generalised this method for arbitrary constitutive models, which required the knowledge of two sets of constitutive parameters for dry and saturated conditions. However, numerous experiments have invalidated the implicit assumption that the sequence of loading and wetting does not affect the final state of materials and led to several attempts to establish other approaches [22–25]. Shen and Wang [22] expressed the magnitudes of the wetting-induced volumetric strain and deviatoric strain using empirical functions, including both confining pressure and deviatoric stress level during the wetting process. By virtue of the assumption of coaxiality of the strain rate tensor and the stress tensor, the strain components are estimated and translated into nodal forces, which are further applied to the finite elements to obtain the wetting deformation of the dam. Recently, this so-called single curve method is widely used in dam engineering in China due to its practicality. Nevertheless, the point to underline here, in the above two procedures, is essentially a computational device that is not necessarily associated with a physical reality [26, 27]. The mechanical properties of rockfill materials are closely related to the breakage properties of rock particles [4, 28] and these breakage properties are sometimes remarkably affected by the degree of saturation [1, 10, 29]. Hence, the physical mechanism underlying the collapse deformation seems to be the difference of breakage rates between saturated and unsaturated rock particles. Therefore, introducing the effect of relative humidity on the rockfill performance is by no means an academic one to model the collapse deformation. Oldecop and Alonso [1] first proposed an elastoplastic model that was valid to express compression properties of unsaturated rockfills considering the influence of relative humidity. Thereafter, within the framework of the Barcelona Basic Model [30] for unsaturated soils, they extended this compression model to the three-dimensional stress state and applied it in the analyses of Beliche and Lechago rockfill dams [31, 32]. Furthermore, Chávez and Alonso [7] developed a work-hardening model to describe the effect of particle breakage at different relative humidities. Besides, Kohgo et al. [33] proposed a three-dimensional elastoplastic model for unsaturated rockfills according to a modification of their generalised elastoplastic model for unsaturated soils. Bauer [34] extended his hypoplastic model by introducing a moisture-dependent solid hardness and taking the stress relaxation caused by the change of solid hardness into consideration. The state-dependent behaviour was not considered in these models.

On the other hand, due to the obvious state-dependent behaviour of sands, many state-dependent models have been proposed [35–45]. Among them, the state-dependent model proposed by Li and Dafalias [41], with only a set of model parameters, can precisely represent the complex stress-strain relationship of Toyoura sand over a wide range of densities and pressures. Within the framework of this model, Xiao et al. [46] developed a generalised elastoplastic model to describe the state-dependent dilatancy for rockfill materials. Sun et al. [47] introduced the fractional order method to describe the state-dependent behaviour of granular materials without using the plastic potential function. However, to the best of our knowledge, these models cannot describe the combination effect of the initial dense state and prevailing relative humidity on the stress-strain relationship of rockfill materials.

The purpose of this chapter is to present an elastoplastic model for unsaturated rockfill materials by introducing a unified hardening parameter which could well capture the essential physical phenomena behind the observed collapse deformation and state-dependent dilatancy for saturated and unsaturated rockfill materials. The conceptual bases for the model and the basic formulation are given in Section 2. Thereafter, the proposed model is incorporated into the coupled flow-deformation analysis FE code [48–51] using the explicit stress integration algorithm for the elastoplastic model of unsaturated soils [52, 53]. Finally, the comparisons are conducted using the experimental data from a series of triaxial tests on saturated and unsaturated rockfill materials, and the capability of the proposed model to capture the state-dependent behaviour and collapse deformation is validated.

2. Constitutive framework and formulations

2.1 Basic constitutive variables

Due to the large permeability for rockfills, water never fills the large rockfill voids unless structures are submerged. Environmental changes and rainfall can only modify the relative humidity. The extreme case of rockfill flooding is a concern in the rockfill structures which become inundated, such as the upstream rockfill shells of the zoned earth and rockfill dams. Hence, a suitable stress space to describe the isotropic compression states of unsaturated rockfill materials is (p, RH), where p is the mean total stress and RH is the relative humidity. In fact, the relative humidity is the ratio of the vapour pressure present in the air to the vapour pressure when the air is saturated with water vapour. The relative humidity in the gas phase and the matric suction, s, in the rockfill pore water are related by the psychrometric relationship [52]:

$$\psi = s + \pi = -\frac{RT\rho_w}{M_w}\ln(RH) \tag{1}$$

where R is the gas constant, T is the absolute temperature of the reference system, M_w is the molecular mass of water and ρ_w is the density of water at the reference temperature. π is called the osmotic suction, which is due to the presence of solutes in the rockfill pore water. The sum $\psi = s + \pi$ is called the total suction. Total and matric suction would be equal with each other in the case when the rockfill pores contained only pure water with no solutes, for instance, rockfill materials applied in the dams. The total stress and total suction are chosen as the basic constitutive variables herein.

2.2 Normal compression model

Considering an isotropic compression test on unsaturated rockfills, in which a rockfill sample at a given relative humidity (subsequently maintained) is subjected to p-load increments along virgin states, which is in full correspondence with the compression behaviour of saturated rockfills, the specific volume will be given by

$$v = 1 + e = N - \lambda(\psi)p \tag{2}$$

where ν is the specific volume, e is the void ratio and N is the intercept of the normal compression lines with the ν-axis when $p = 0$. The compression index $\lambda(\psi)$ is assumed to be a function of the total suction and can be interpolated from the compression indices at the fully saturated state ($\psi = 0$) and very dry state. For rockfill materials, the very dry state is usually defined as the water content close to zero, generally corresponding to the water content less than 0.45% in practice. A schematic representation of Eq. (1) can be discretized into a series of normal compression lines, as shown in **Figure 1**.

As mentioned by McDowell and Bolton [53], the plastic deformation of the granular materials can be attributed to the two deformation mechanisms. Under low stress, the plastic deformation is due to the particle rearrangement. The second mechanism is called clastic yielding, and it is attained when the applied stress causes the onset of particle breakage. According to this concept and corresponding experimental results, an elastoplastic compressibility model for rockfill was developed by Oldecop and Alonso [1]. Under isotropic compression, a mean threshold total stress, p_y, which marks the beginning of particle breakage, was introduced in the model. Below this threshold stress, the influences of suction on the compression index were

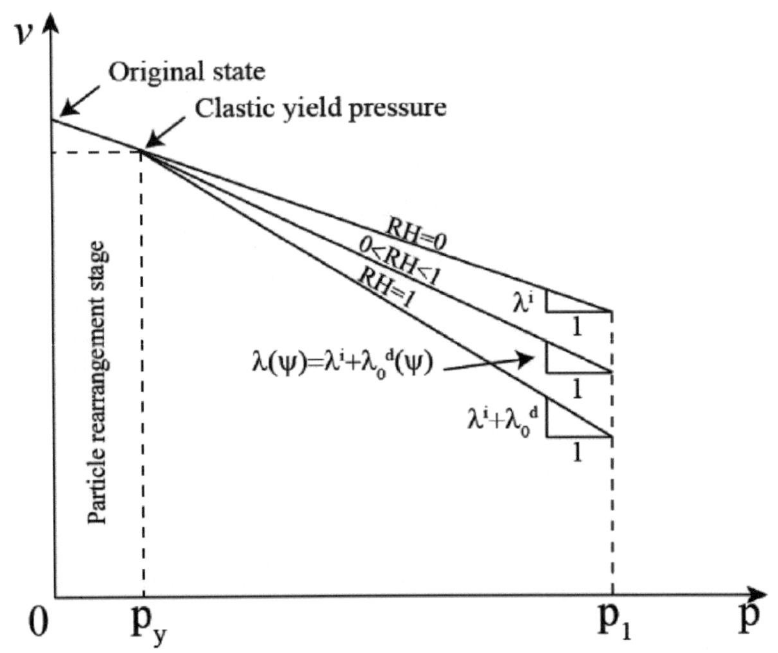

Figure 1.
Normal compression lines for saturated and unsaturated rockfill materials.

not measured in the compression tests. On the other hand, the effect of water action on the compression behaviour was strong when beyond this threshold stress, which is the source of the collapse deformation. Note that, the difference of compression index below and beyond this threshold stress is vanished when the material is in the very dry state. Therefore, the compression index for unsaturated rockfills can be expressed as

$$\begin{array}{ll} p \leq p_y & \lambda(\psi) = \lambda^i \\ p > p_y & \lambda(\psi) = \lambda^i + \lambda^d(\psi) \end{array} \quad (3)$$

where λ^i is the slope of the normal compression line when particle rearrangement is active only, and $(\lambda^i + \lambda^d)$ is the slope of the normal compression line when both plastic deformation mechanisms (particle rearrangement and particle breakage) are active. According to the previous literature [1], the compressibility index accounting for the clastic yielding is linearly related to the logarithm of total suction through a material parameter:

$$\lambda^d(\psi) = \lambda_0^d - \alpha_\psi \ln\left(\frac{\psi + p_{atm}}{p_{atm}}\right) \quad (4)$$

where λ_0^d is the maximum clastic compressibility index in the saturated condition, p_{atm} is the atmospheric pressure and α_ψ is a model parameter. The elastic volume change for rockfill materials within unloading-reloading (URL) paths are given by

$$dv = -\kappa dp \quad (5)$$

where the slope κ is assumed to be independent of the water action. In addition, there was a moderate swelling behaviour observed in the experiments on the unsaturated rockfills, which was caused by the increase in water content. This swelling is assumed to be a reversible elastic deformation, and a linear relationship is assumed between the swelling volume change and the logarithm of total suction:

$$dv^\psi = -\kappa_\psi \frac{d\psi}{(p_{atm} + \psi)} \quad (6)$$

where κ_ψ is the total suction-based expansion/compression index. For simplicity, κ_ψ is assumed to be independent of the stress level.

2.3 Loading-collapse yield curve

For the isotropic compression condition, the yield stress, p_0, was defined by the following expression such as [1]:

$$\begin{cases} p_0^* \leq p_y \Rightarrow p_0(\psi) = p_0^* \\ p_0^* > p_y \Rightarrow p_0(\psi) = p_y + \dfrac{(\lambda^i - \kappa)\left(p_0^* - p_y\right)}{\lambda^i + \lambda^d(\psi) - \kappa} \end{cases} \quad (7)$$

Eq. (7), which is referred to as loading-collapse (LC) yield surface, describes the relationship between the equivalent yield stress (p_0^* defines the position of the yield curve, and it was identified as the yield stress for a very dry rockfill) and the yield

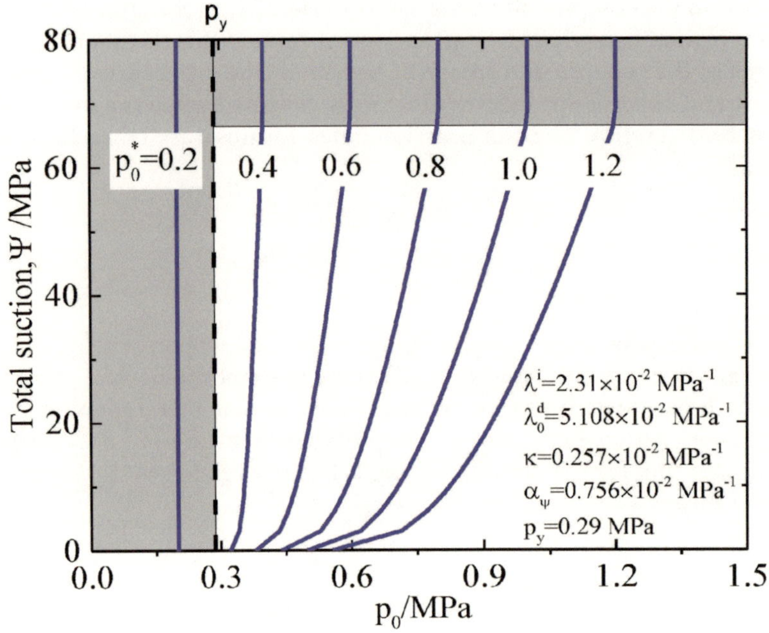

Figure 2.
Load-collapse curves for different values of p_0^*.

stress (p_0) in unsaturated states, which is controlled by the total suction. By introducing the compressibility index for the clastic yielding expressed in Eq. (4) into Eq. (7), the LC yield surface is then obtained. A plot of the LC yield curves for different values of p_0^* for Pancrudo slate that were tested by Oldecop and Alonso [1] is given in **Figure 2**. There are two regions in which the yield stress, p_0, does not depend on the total suction, which correspond to the stress levels below the clastic yield stress py ($p_0^* < 0.29$ MPa) and to the very dry states ($\psi > 67$ MPa), respectively.

Thereafter, a simple volumetric hardening is used here to follow the evolution of p_0^*:

$$dp_0^* = \frac{dH}{\lambda^i - \kappa} \tag{8}$$

where H is the unified hardening parameter related to the plastic volumetric strain and state-dependent behaviour, which will be introduced in Section 2.5.

2.4 Yield surface

Yao et al. [54] developed a simple but robust constitutive model (the original UH model) to reproduce the mechanical behaviours in the isotropic and triaxial states for both normally consolidated (NC) and overconsolidated (OC) soils, such as non-elastic deformation in reloading, peak strength, shear-dilation and strain-softening. In his model, a unified hardening parameter (H) was proposed, and an associated flow rule was adopted for the robust and convenience of numerical implementation. Hence, in accordance with the aim of simplicity, the original UH model is adopted as the base for developing a new state-dependent constitutive model of unsaturated rockfill materials. Thus, the yield and plastic potential surfaces in the stress space (p, q and ψ) are defined as follows:

$$f = g = \begin{cases} p + \dfrac{q^2}{M^2 p} - \dfrac{H}{\lambda^i - \kappa} & if\, p_0^* \leq p_y \\ p + \dfrac{q^2}{M^2 p} - \dfrac{\lambda^d(\psi)p_y - H}{\lambda^i + \lambda^d(\psi) - \kappa} & if\, p_0^* > p_y \end{cases} \tag{9}$$

where q is the deviator stress and M is the critical-state slope.

With respect to the rockfill materials, the critical-state slope, M, is no longer a constant. Experimental results on rockfill materials show that the critical state friction angle is dependent on the mean stress [19, 55–57] that decreases with an increase in p. Moreover, according to the RH-controlled triaxial tests on rockfill materials in Chávez and Alonso [7], M was also found to be dependent on the RH, and the relationship between M and $\log\sigma_3$ for samples with different RH seemed to be a series of parallel lines. From the experimental data [7, 16], the value of M increases with the increase of total suction (decrease RH). Besides, considering the critical stress ratio is impossible to be negative with the increase of mean stress, the following function is adopted for the sake of simplicity:

$$M(p,\psi) = \left[M_{res} + (M_0 - M_{res})e^{-\alpha_M p}\right]\left[1 + \beta_\psi \ln\left(\frac{\psi + p_{atm}}{p_{atm}}\right)\right] \tag{10}$$

where M_o and M_{res} are the initial and residual stress ratios for rockfills in saturated condition under low (0.1 MPa) and high confining pressures, respectively; α_M and β_ψ are material constants. **Figure 3** shows that Eq. (10) can well describe the change of the critical stress ratio with mean pressure and relative humidity for unsaturated rockfill materials.

Using above equations, a three-dimensional view of the yield surfaces in the (p, q and ψ) space is given in **Figure 4**, where ψ_i is the total suction corresponding to the very dry state.

2.5 Unified hardening parameter

The unified hardening parameter was first proposed to provide a unified description of the mechanical behaviours for both clays and sands by Yao et al. [58]. It was then modified and employed to model the hardening process for both normally consolidated and overconsolidated clays [54, 59]. The unified hardening parameter, which is adopted to describe the hardening of the yield surface of rockfill materials can be written as

$$H = \int dH = \int \frac{M_f^4 - \eta^4}{M^4 - \eta^4} d\varepsilon_v^p \tag{11}$$

where η is the stress ratio, $\eta = q/p$; ε_v^p is the plastic volumetric strain; and M_f is the potential failure stress ratio, which represents the potential peak strength that changes with the compaction density and relative humidity. Since dH is always larger than or equal to zero, the following conclusions can be drawn from Eq. (11):

1. $0 \leq \eta < M$ (negative dilatancy condition): $d\varepsilon_v^p > 0$.

2. $\eta = M$ (characteristic state condition): $d\varepsilon_v^p = 0$.

3. $M \leq \eta < M_f$ (positive dilatancy condition): $d\varepsilon_v^p < 0$.

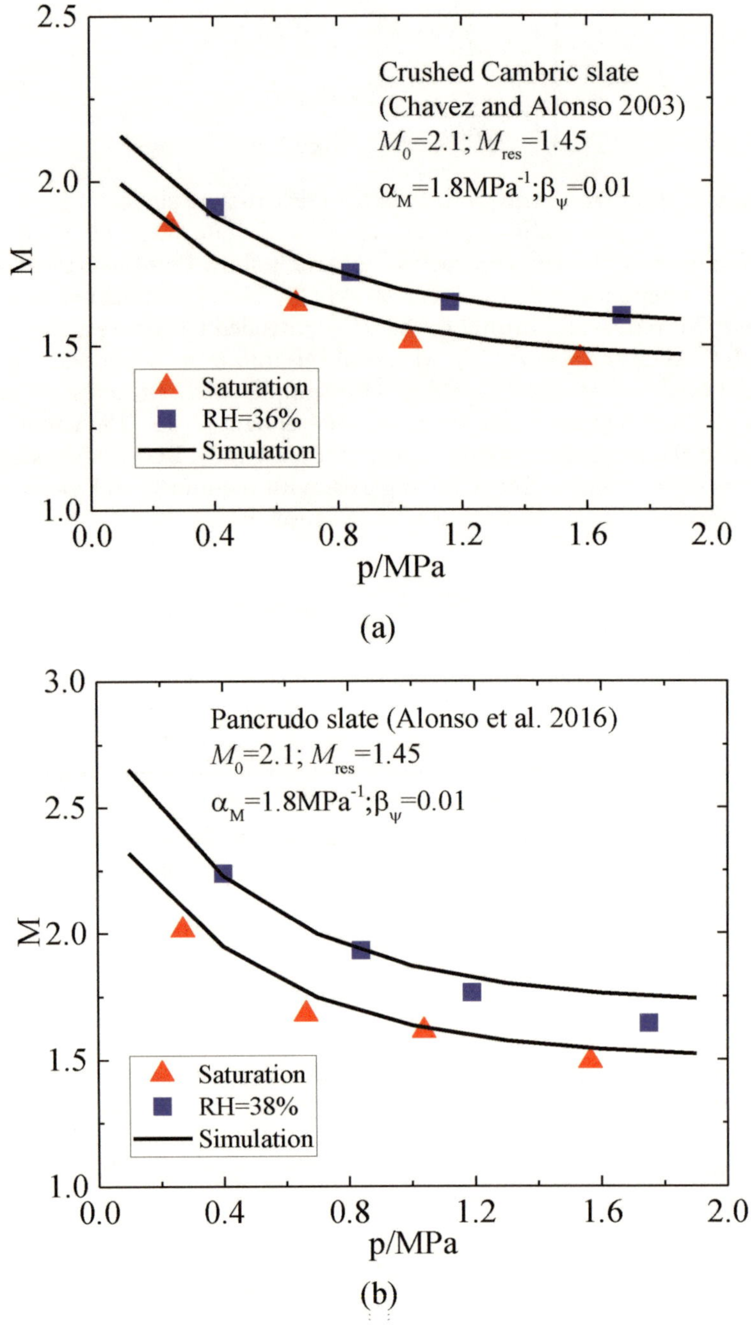

Figure 3.
Variations of critical stress ratio with mean stress and relative humidity: (a) crushed cambric slate and (b) Pancrudo slate.

As indicated above, the dilatancy of rockfill materials can be reasonably described by this hardening parameter.

The next is introducing the determination of potential failure stress ratio. For rockfill materials, the peak friction angle, ϕ_p, on a contact plane is dependent on the degree of interlocking by neighbouring particles, which can be related to the state of the packing void ratio [60]:

$$\tan\phi_p = \left(\frac{e_c}{e}\right)^m \tan\phi_\mu \tag{12}$$

where ϕ_μ is the internal friction angle corresponding to the critical state, which has the relationship with the critical stress ratio M as shown in Eq. (17); and m is a material constant when granular materials are in the fully saturated condition, but it varies with the confining pressure and suction in unsaturated conditions according to the experimental observations, which can be corrected using the following function:

$$m(\sigma_3, \psi) = m_2 + (m_1 - m_2)\left(\frac{m_2}{m_1}\right)^{\psi\left(\frac{\sigma_3}{p_{atm}^2}\right)^n} \tag{13}$$

where σ_3 is the confining pressure; m_1, m_2 and n are material constants. From Eq. (13), we can find that parameter m is kept as m_1 in the saturated condition, but decreases with the increasing confining pressure when in the unsaturated condition. Thus, the nonlinear variation of the peak strength with the changes of suction and confining pressure can be well described using this function.

Furthermore, e_c in Eq. (12) is the void ratio corresponding to the critical state, which is a function of the mean stress. However, for the rockfills in the unsaturated condition, the influence of the total suction on the critical void ratio should be considered as well. From the experimental data in [7, 16], the relation between critical stress ratio and mean stress for saturated and unsaturated rockfills seems to be two parallel lines. Thus, the critical void ratio can be expressed based as [61].

$$e_c(p, \psi) = e_{ref}(\psi) - \zeta \log\left(\frac{p}{p_{ref}}\right) \tag{14}$$

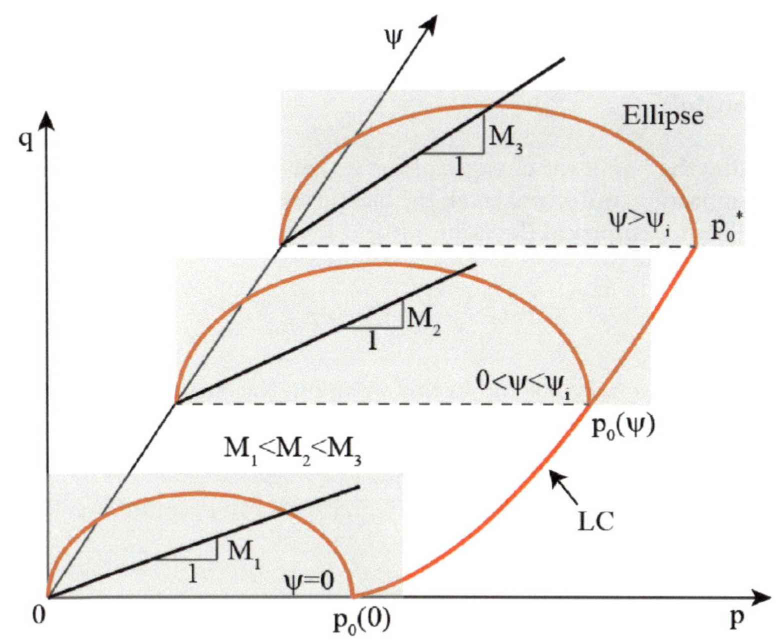

Figure 4.
Three-dimensional yield surface in stress space (p, q and ψ).

where ζ is the material constant; (e_{ref}, p_{ref}) is a reference point on the critical state line e_{ref} is a function of total suction, which nonlinearly increases with the increase of total suction. For the sake of simplicity, the following function is used to express the variation of reference critical void ratio with total suction:

$$e_{ref}(\psi) = e_{ref}^0 + \alpha_e \ln\left(\frac{\psi + p_{atm}}{p_{atm}}\right) \tag{15}$$

where e_{ref}^0 means the reference critical void ratio in the saturated condition; and α_e is the material constant, which describes the increase rate of reference critical void ratio with total suction.

In the triaxial compression condition, the stress ratios, M_f and M, can be expressed by using peak friction angle (ϕ_p) and internal friction angle (ϕ_μ) as

$$M_f = \frac{6 \sin \phi_p}{3 - \sin \phi_p} \tag{16}$$

$$M = \frac{6 \sin \phi_\mu}{3 - \sin \phi_\mu} \tag{17}$$

Combining Eqs. (12), (16) and (17), the relation between the potential stress ratio (M_f) and critical state stress ratio (M) can then be obtained.

Substituting Eq. (16) into Eq. (11), the hardening parameter is then obtained, which can reflect the influence of the compacted density and relative humidity on the mechanical behaviour of unsaturated rockfill materials. Considering the peak frictional angle, ϕ_p, is generally greater than ϕ_μ, especially in the dense and very dry states, the dilatancy behaviour for the rockfill materials can thus be well captured using the modified hardening parameter. In addition, once an obvious dilatancy occurs for the rockfill in the low confining pressure, the degree of interlocking and the peak frictional angle are reduced, and the strain-softening phenomenon can also be well described by using this hardening parameter.

2.6 Elastic moduli

Considering the nonlinear changes of the elastic moduli with the changes of degree of compaction and stress level, the elastic shear modulus, G, is calculated by using the following empirical Equation [46]:

$$G = G_0 \frac{(2.97 - e)^2}{1 + e} (pp_{atm})^{0.5} \tag{18}$$

where G_0 is a material constant. In this equation, the current void ratio, e, is used instead of the initial void ratio. The elastic bulk modulus, K, is equal to

$$K = \frac{2(1 + \nu)}{3(1 - 2\nu)} G \tag{19}$$

where ν is the Possion's ratio.

2.7 Determination of model parameters

The application of the model requires the information on the following stress states and parameters.

a. Initial state: initial stresses (p_i, q_i and s_i), initial void ratio, e_0, and initial mean yield stress, p_0^*.

b. Parameters directly associated with the isotropic compression behaviour: p_y, threshold yield mean stress for the onset of clastic phenomena; λ^i, normal compressibility index when instantaneous deformation mechanism is active only; λ_0^d, maximum clastic compressibility index for fully saturated conditions; κ, compressibility coefficient along elastic (unloading-reloading) stress paths and α_s, compressibility parameter, which controls the rate of stiffness increase with total suction.

c. Parameters directly associated with the critical state: M_0, maximum critical stress ratio for the mean stress approaching to 0.1 MPa; M_{res}, residual critical stress ratio corresponding to the large mean stress; α_M, controls the rate of critical stress ratio decrease with mean stress; β_ψ, controls the rate of critical stress ratio increase with total suction; (e_{ref}^0, p_{ref}) is a reference point on the critical state line in the saturated conditions; χ, controls the rate of critical void ratio decrease with mean stress and α_e, controls the rate of critical void ratio increase with total suction.

d. Parameters directly associated with elastic moduli: G_0, material constant associated with shear modulus within the elastic domain; ν, Possion's ratio, which is assumed to be independent of suction and taken as $0.2 \sim 0.3$ for rockfill materials and κ_ψ, compressibility coefficient for changes in suction within the elastic region.

e. Other default parameters: m_1, controls the rate of peak strength change with void ratio in the saturated conditions; m_2, minimum value of parameter, m, corresponding to the high confining stress at very dry state and n, controls the vary of m with suction and confining pressure.

In general, the determination of the model parameters will require relative humidity-controlled testing methods, and suggested stress paths for different sets of parameters are the following:

a. Tests that involve isotropic compression (loading and unloading) for very dry specimen and saturated ones; they provide data to find λ^i, p_0^*, λ_0^d and κ.

b. Tests that involve isotropic compression for very dry specimen subsequent with flooding under low applied stress and constant stress p_0; they provide data to find κ_ψ and p_y. As mentioned in the previous study [1], p_y can be obtained using the following function:

$$p_y = p_0 - \frac{\varepsilon_{collapse} + \varepsilon_{expansion}}{\lambda_0^d} \tag{20}$$

where $\varepsilon_{expansion}$ is the measured total expansion strain due to flooding under low applied stress and $\varepsilon_{collapse}$ is the measured total collapse strain due to flooding under applied stress p_0.

c. Tests that involve isotropic compression under different relative humidity from what a linear relationship between collapse strains and the logarithm of rockfill water content can be observed; Oldecop and Alonso [1] provided a method to find α_ψ using the following function

$$\alpha_\psi = \frac{\kappa_\psi + \chi_\psi}{p_0 - p_y} \tag{21}$$

where χ_ψ is the experimental coefficient relating collapse strain with total suction at constant stress p_0 ($p_0 > p_y$).

d. Triaxial tests (loading, unloading and reloading) at different relative humidities and saturated condition, and they provide data to find M_0, M_{res}, α_M, β_ψ, e^0_{ref}, p_{ref}, χ, α_e and G_0.

Note that the default material parameters, m_1, m_2 and n, should be determined according to the back analysis using the experimental data by fixing the previous determined parameters. Besides, the above-mentioned test is the minimum experimental programme, so the more tests are required to determine the more reliably model parameters.

3. Model validations

In order to verify the capability and feasibility of the proposed constitutive model, it was incorporated into the hydro-mechanical coupling FEM code [48] that was developed in the Institute of Hydraulic Structures of Hohai University [49, 50]. The explicit stress-integration algorithm for unsaturated soils proposed by Sheng et al. [51, 62] was used for the numerical implementation of the proposed constitutive model. The triaxial tests on the saturated specimens with different initial void ratio, triaxial tests with different relative humidity and triaxial wetting tests under constant vertical strain rate were simulated by means of a hexahedral element. The confining pressure was applied against the upper and lateral boundaries, a constant vertical displacement rate was then applied to the upper boundary for shearing and the axial stresses and volumetric strains were computed finally. For the tests at dry state, a constant high suction, which represents the laboratory conditions, was maintained in the specimen. Flooding was simulated by converting negative pore-water pressures (total suctions) of element nodes into zero.

3.1 Behaviour of rockfill compacted to various initial void ratios

At first, the model was used to predict drained triaxial tests on saturated rockfills with different initial void ratios. The mixture of weak and strong weathered limestone, which was selected as the rockfill material for the Guanyinyan dam (Panzhihua, China), was used in the triaxial tests. The gradation of the tested material was scaled down by making the specimen's cumulative grain size distributions parallel to the gradation curve of the in situ rockfill material combining with the equivalent substitution method. **Figure 5** shows the grain size distribution. The initial void ratio was controlled as 0.24, 0.28 and 0.37 in the sample preparation, respectively. Before shearing, the samples were saturated and isotropically compressed to the confining pressures ranging from 0.2 to 1.2 MPa. The model parameters, which are listed in **Table 1**, were determined using the procedure introduced in the above section based on the experimental stress-strain relationships. Note that, rockfill specimens for laboratory tests were prepared by compacting the material in

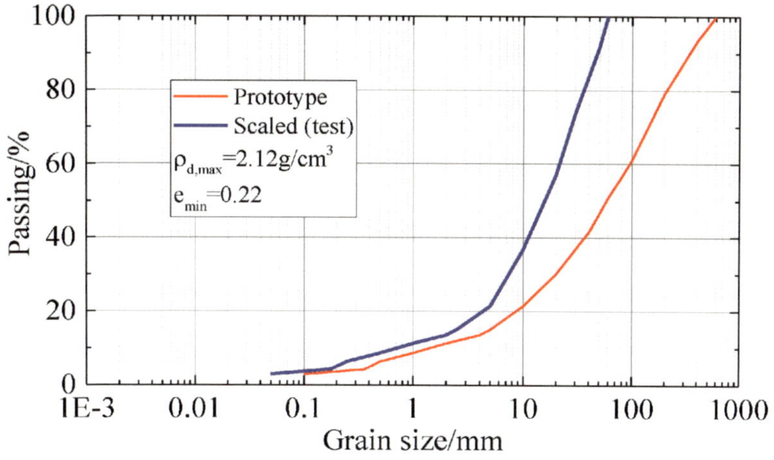

Figure 5.
Grain size distribution curves.

Elastic parameters	Isotropic compression parameters	Critical state parameters	Default parameters
G_0 = 125	λ^i-κ = 0.001 MPa^{-1}	M_0 = 1.95	m_1 = 0.36
ν = 0.28	λ_0^d = 0.02 MPa^{-1}	M_{res} = 1.55	m_2 = 0.1
κ_ψ = 0	α_ψ = 0.001 MPa^{-1}	α_M = 1.2 MPa^{-1}	n = 3.5
	p_y = 0.01	β_ψ = 0.01	
	p_0^* = 0.3 MPa	e_{ref}^0 = 0.51	
	e_0 = 0.24, 0.28 and 0.37	p_{ref} = 0.02 MPa	
		ζ = 0.105	
		α_e = 0.03	

Table 1.
Model parameters for tested limestone.

thin layers to obtain uniform samples [63]. Thus, a significant pre-consolidation effective mean stress (p_0^* = 0.3 MPa), associated with compaction, was used in the model to simulate the triaxial tests.

The comparisons between the numerical results and experimental measurements at different confining pressures are shown in **Figure 6**. It can be found that proposed model is satisfied to capture the key features of the loose and dense saturated specimens with one unique set of soil parameters. For instance, the denser of the specimen, the larger the deformation modulus and shear strength. Besides, it clearly shows that the rockfill with a smaller initial void ratio (denser state) dilates more remarkably due to the difference between the peak stress ratio and critical stress ratio is much greater, especially at a low confining pressure. With the increase of initial void ratio, the difference between the peak stress ratio and critical stress ratio decreases, which leads to the decrease of volumetric dilatancy degree. It is worth to note that there is an obvious strain-softening phenomenon observed for the dense specimen under the low confining pressures. To our satisfaction, the proposed model can also predict this strain-softening to some extent, although there is a little difference which may be caused by the difficult to determine the parameters corresponding to the critical state.

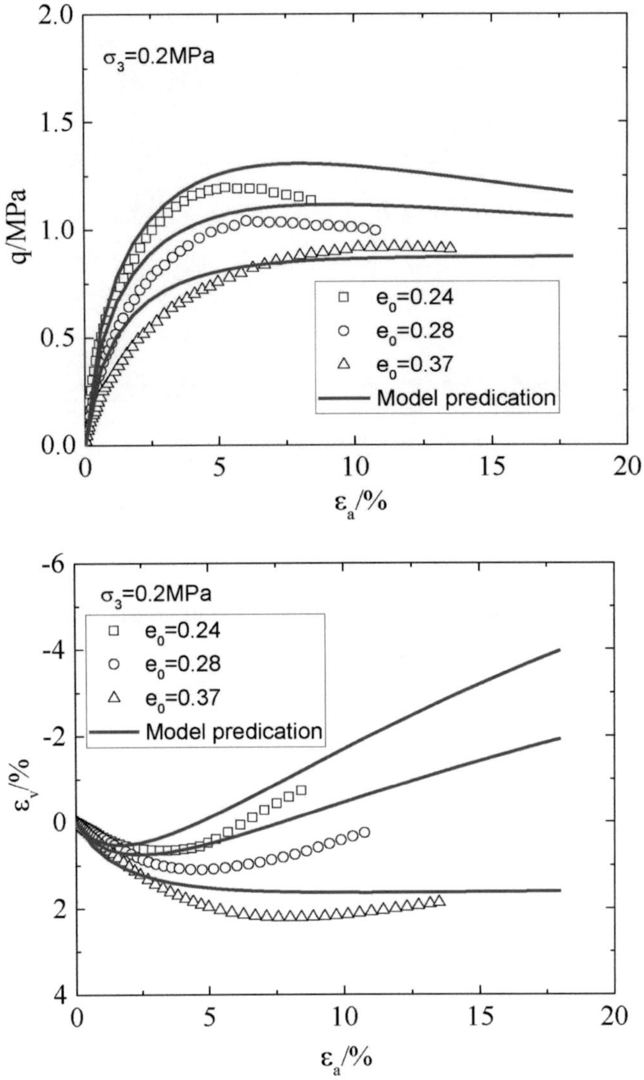
σ3=0.2MPa
q/MPa
2.0
1.5
1.0
0.5
0.0
0
5
10
15
20
εa/%
e0=0.24
e0=0.28
e0=0.37
Model predication
σ3=0.2MPa
e0=0.24
e0=0.28
e0=0.37
Model predication
εv/%
-6
-4
-2
0
2
4
0
5
10
15
20
εa/%

(a)

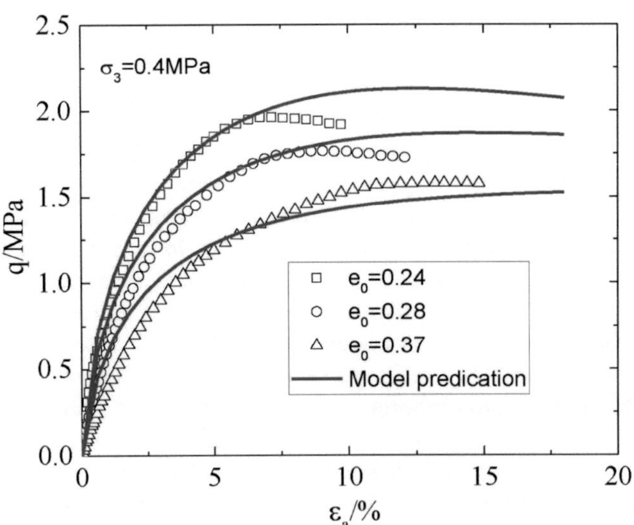
σ3=0.4MPa
q/MPa
2.5
2.0
1.5
1.0
0.5
0.0
0
5
10
15
20
εa/%
e0=0.24
e0=0.28
e0=0.37
Model predication

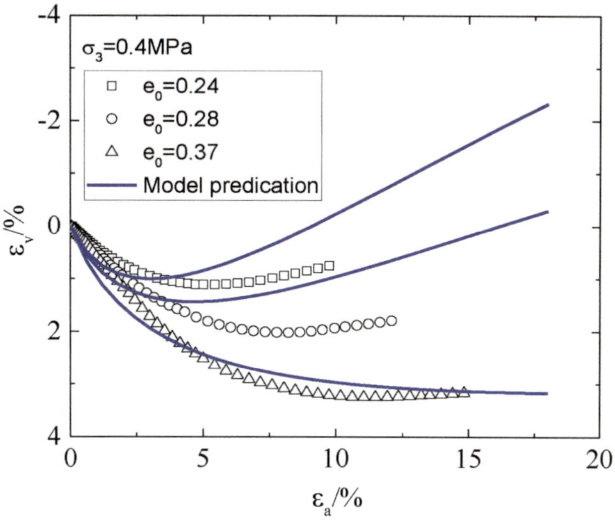
σ3=0.4MPa
e0=0.24
e0=0.28
e0=0.37
Model predication
εv/%
εa/%
-4
-2
0
2
4
0
5
10
15
20

(b)

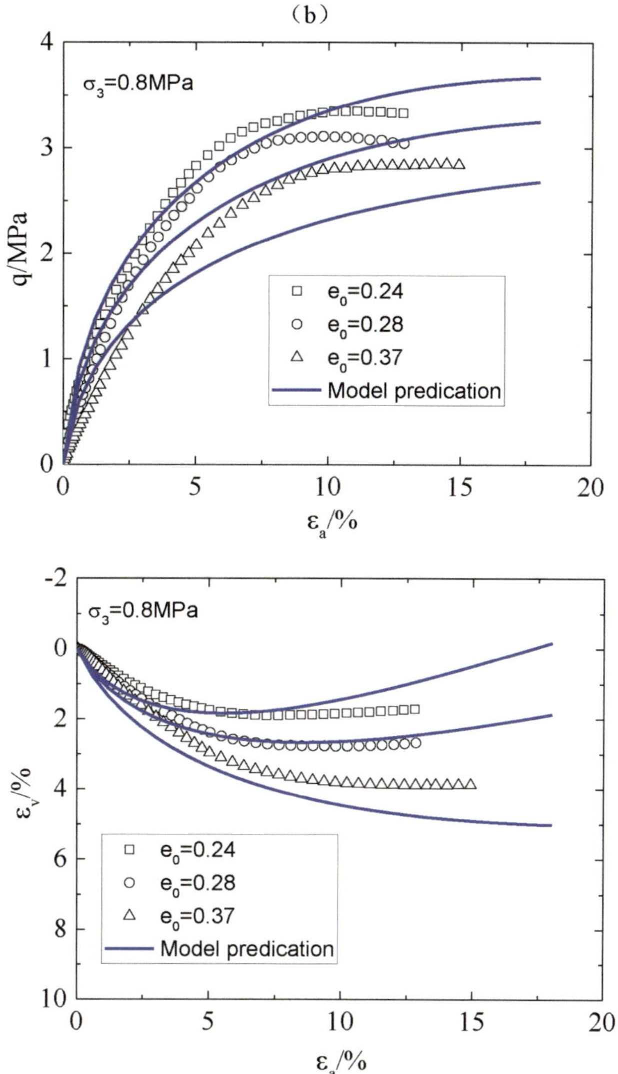
σ3=0.8MPa
q/MPa
εa/%
e0=0.24
e0=0.28
e0=0.37
Model predication
0
1
2
3
4
5
10
15
20
σ3=0.8MPa
εv/%
εa/%
e0=0.24
e0=0.28
e0=0.37
Model predication
-2
0
2
4
6
8
10
5
10
15
20

(c)

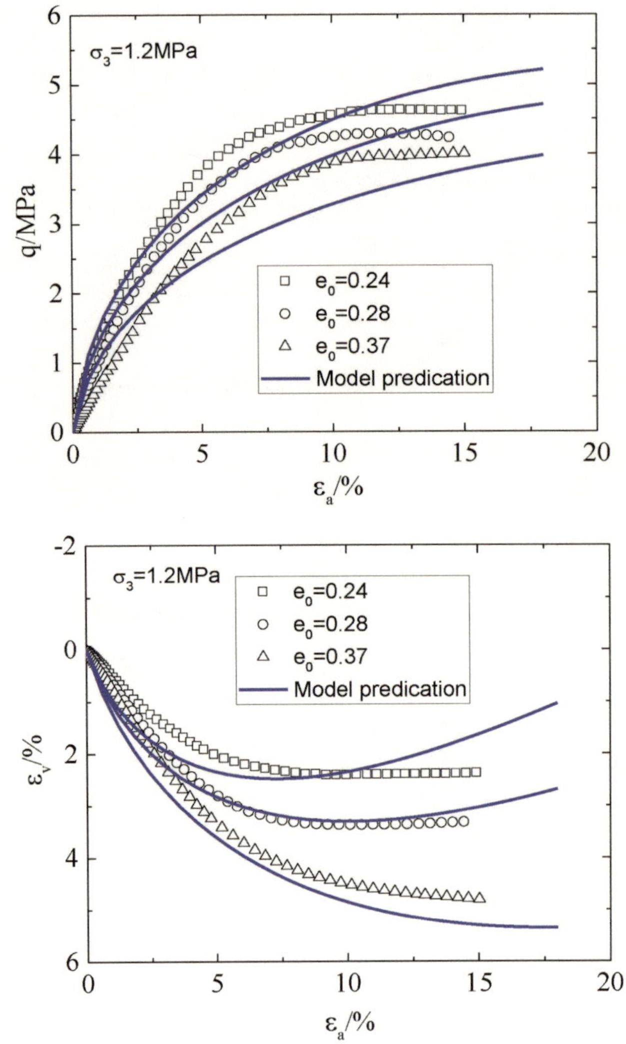

(d)

Figure 6.
Measured and predicted stress-strain relationship (q vs. εa and εv vs. εa) for rockfill with different initial void ratios under confining pressures: (a) σ_3 = 0.2 MPa, (b) σ_3 = 0.4 MPa, (c) σ_3 = 0.8 MPa and (d) σ_3 = 1.2 Mpa.

Figure 7 shows the predicted relationship between the potential failure stress ratio, M_f, and stress ratio, η, for three different initial void ratios when σ_3 = 0.2 MPa, which can be used to explain the strain-softening phenomenon. From the initial state to the final state, the state $\eta = M_f$ (i.e., the peak strength state) appears, shown as the intersections of the 45^0 line and the M_f (η) lines for different initial void ratios. Before the peak strength state is reached, $\eta < M_f$, the rockfill material undergoes a hardening process and remains in a stable state. After the peak strength state, η becomes slightly larger than M_f, and then the rockfill material undergoes a softening process that is unstable. It can be found that the part of $\eta > M_f$ becomes more remarkable with the decrease of the initial void ratio, which explains the appearance of the most significant strain-softening process for the densest specimen (e_0 = 0.24). Additionally, **Figure 8** presents the influence of the confining pressure on the changes of potential failure stress ratio with the changes of stress

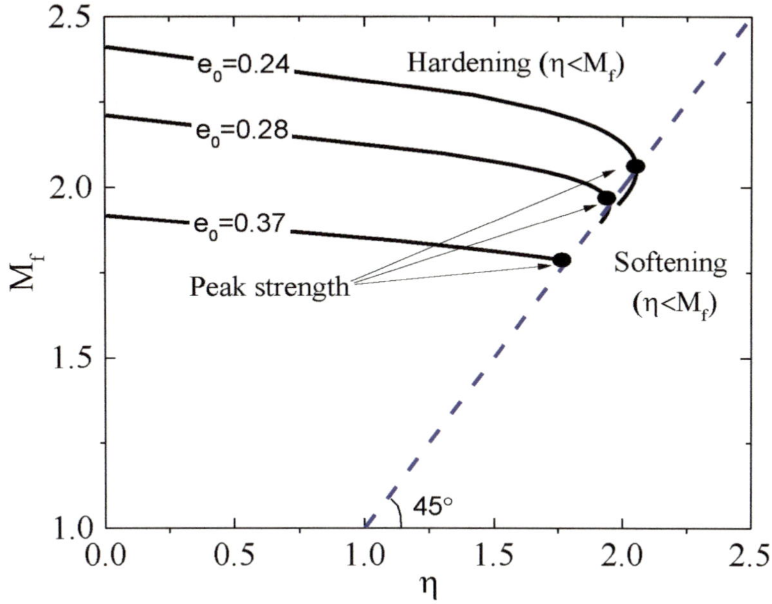

Figure 7.
Relationships between potential failure stress ratio, mf, and stress ratio, η, for three different initial void ratios when σ_3 = 0.2 MPa.

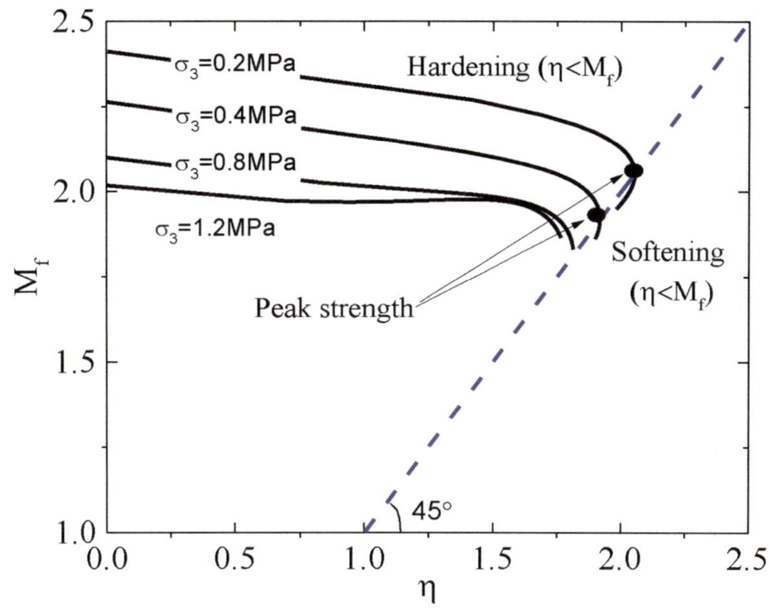

Figure 8.
Relationships between potential failure stress ratio, mf, and stress ratio, η, for different confining pressures when e_o = *0.24.*

ratio when e_0 = 0.24. It can be found from this figure that M_f (η) line for the larger confining pressure is always in the left side of the 45^0 line ($\eta < M_f$) from the initial state to the final state. Thus, there is no softening phenomenon appearing when there is a large confining pressure applied on the rockfills, and the deviatoric stress only increases up to the strength corresponding to the critical state.

Noting that, the difference of stress-strain curve between the measured and predicted is relatively larger with the increase of confining pressure, which is shown in **Figure 6(c)** and **(d)**. Compared with that experimental stress-strain response, the predicted deviatoric stress increases more slowly with the increase of axial strain. From our analysis, this may be caused by the significant particle breakage under the high confining pressures for the rockfills used in the tests, which led to the considerable increase of compressibility modulus before shearing. However, the influence of the particle breakage on the compressibility modulus of rockfill materials has not been considered in the proposed model. In the construction of model, a linear relationship in the e-p plane is assumed, which means the compression index is kept unchanged throughout the isotropic and triaxial compression process. Even so, the quantitative results are in general in good agreement with the experimental measurements. Finally, from above comparisons, it is indicated that the behaviours of the state-dependent dilatancy and strain-softening for rockfill materials can be well captured by the proposed model.

3.2 Behaviour of rockfill with various RH

In order to illustrate the capability of proposed model to predict the stress-strain relationships of unsaturated rockfills, the results of RH-controlled triaxial tests on the crushed cambric slate were adopted as comparison. These tests were conducted by Chavez and Alonso [7], which involved two series of tests. The first series was performed in the fully saturated condition. After compaction, the samples were confined under pressures of 0.1, 0.3, 0.5 and 0.8 MPa and then sheared under the constant confining pressure. In a second series, samples were compacted in a similar manner as saturated ones and maintained at a constant low relative humidity (RH = 36%, equivalent to the total suction, s = 142 MPa). After that, similar stress paths were applied to these dry samples. The model parameters are given in **Table 2**. Most of parameters were taken from the Chavez and Alonso [7], while the rest were determined according to the experimental stress-strain curves.

Figure 9 shows the predicted and measured stress-strain curves in the conditions of fully saturated and dry (RH = 36%). For the saturated sample under a low confining pressure (0.1 MPa), the deviatoric stress increases with the axial strain up to a constant value for axial strains reaching around 10% and then remains constant, which shows the hardening process. However, for the dry sample with the same confining pressure, a peak strength is observed for axial strain reaching the value between 4 and 6%, and a little strain-softening appears after that. On the other hand, when the sample is sheared under the high confining pressures, the

Elastic parameters	Isotropic compression parameters	Critical state parameters	Default parameters
$G_0 = 80$	$\lambda^i\text{-}\kappa = 0.001\ MPa^{-1}$	$M_0 = 2.1$	$m_1 = 0.36$
$\nu = 0.29$	$\lambda_0^d = 7.48 \times 10^{-2}\ MPa^{-1}$	$M_{res} = 1.45$	$m_2 = 0.1$
$\kappa_\psi = 0.4 \times 10^{-3}$	$\alpha_s = 7.86 \times 10^{-3}\ MPa^{-1}$	$\alpha_M = 1.8\ MPa^{-1}$	$n = 3.5$
	$p_y = 0.01\ MPa$	$\beta_\psi = 0.01$	
	$p_0^* = 0.3\ MPa$	$e_{ref}^0 = 1.0$	
	$e_0 = 0.6$	$p_{ref} = 0.01\ MPa$	
		$\zeta = 0.1$	
		$\alpha_e = 0.03$	

Table 2.
Model parameters for crushed cambric slate.

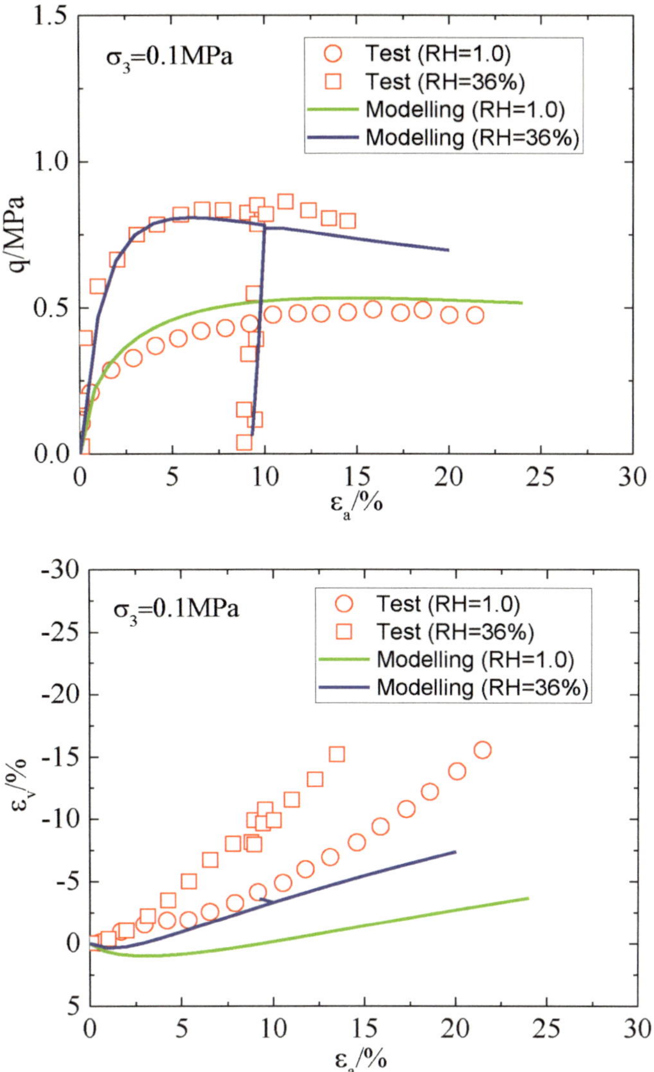
σ3=0.1MPa
Test (RH=1.0)
Test (RH=36%)
Modelling (RH=1.0)
Modelling (RH=36%)
q/MPa
εa/%
σ3=0.1MPa
Test (RH=1.0)
Test (RH=36%)
Modelling (RH=1.0)
Modelling (RH=36%)
εv/%
εa/%

(a)

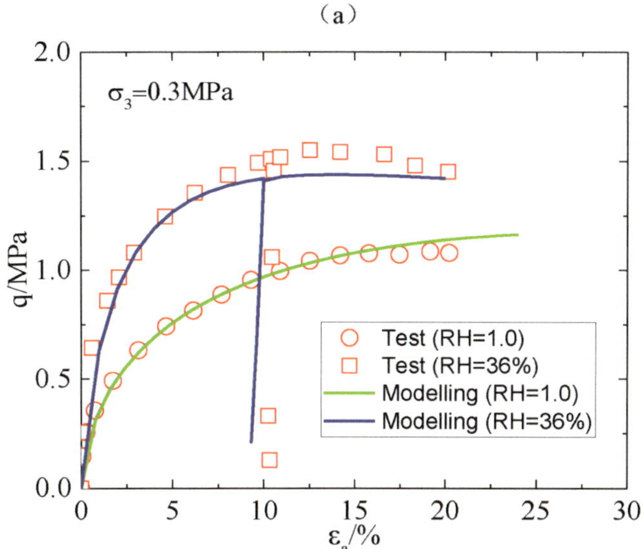
σ3=0.3MPa
Test (RH=1.0)
Test (RH=36%)
Modelling (RH=1.0)
Modelling (RH=36%)
q/MPa
εa/%

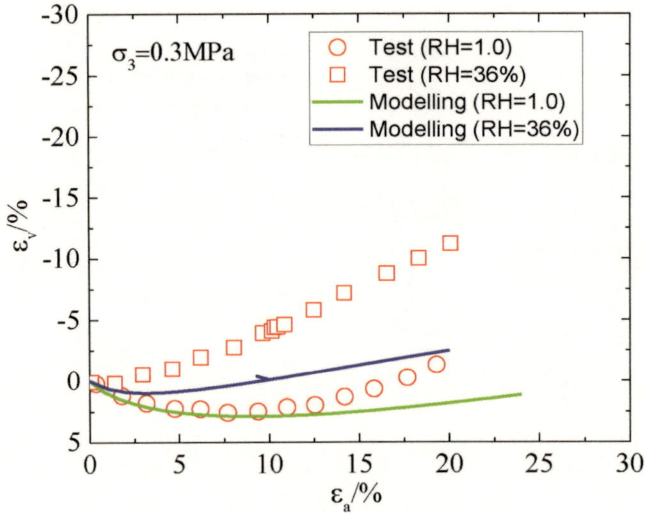
σ3=0.3MPa
Test (RH=1.0)
Test (RH=36%)
Modelling (RH=1.0)
Modelling (RH=36%)
εv/%
εa/%

(b)

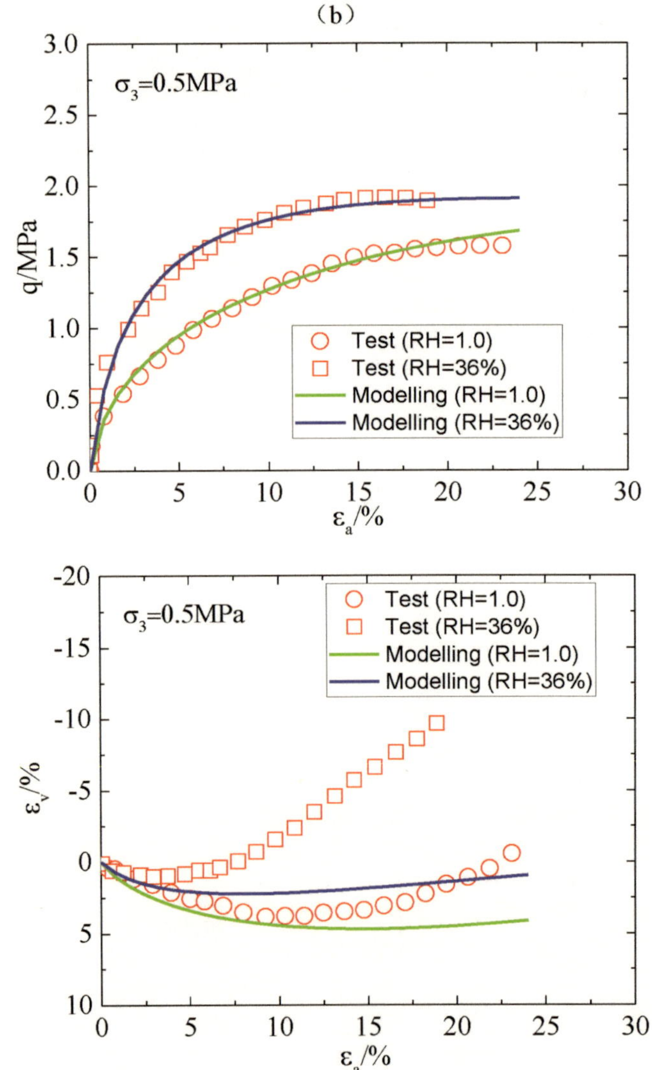
σ3=0.5MPa
q/MPa
εa/%
Test (RH=1.0)
Test (RH=36%)
Modelling (RH=1.0)
Modelling (RH=36%)
σ3=0.5MPa
Test (RH=1.0)
Test (RH=36%)
Modelling (RH=1.0)
Modelling (RH=36%)
εv/%
εa/%

(c)

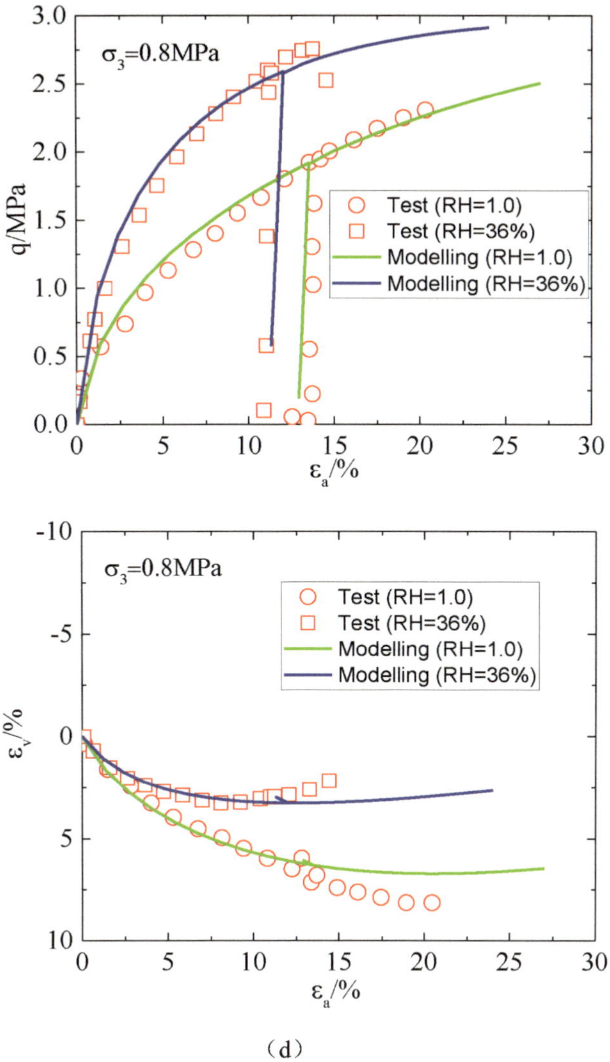

Figure 9.
Comparison between model predications and experimental measurements: Triaxial tests on rockfill in the fully saturated and dry states (RH = 36%). *(a)* σ_3 = 0.1 MPa. *(b)* σ_3 = 0.3 MPa. *(c)* σ_3 = 0.5 MPa. *(d)* σ_3 = 0.8 MPa.

increase of deviatoric stress with the axial strain continues at the end of triaxial tests for both saturated and dry samples. Thus, it is difficult to determine the critical state condition in this case. It can be found a good agreement of deviatoric curves (q vs. ε_a) between modelling and experimental data at both saturated and dry states. Besides, the unloading-reloading process can be well predicted as well with the proposed model. However, some difficulties are observed in reproducing the volumetric response by this model. Computed dilation rates, especially at the low confining pressures, tend to underestimate measured values. Experimental measured dilation rates are significantly larger than that predicted, especially for the unsaturated rockfills under the low confining pressures. There are two reasons accounting for this difference. The first is possibly due to the assumption of critical state conditions in the determination of model parameters. In fact, the attainment of the critical state conditions for a gravel material is probably outside the possibilities of a triaxial test. Very larger strains and a fully broken gravel structure are probably required, but strain localization is also the failure mode in most of tests, which will complicate the interpretation. Thus, the proper observation of critical state conditions would require the larger deformations which were not achieved in the triaxial tests. The second relevant reason is the shape of particles in a gravel. The tested rockfill material has elongated particles with sharp edges, so interlocking and dilation seem very likely to happen even after some significant breakage and deformation, and this also increases the difficulty to modelling the volumetric strains.

3.3 Behaviour of rockfill during shearing and wetting

Finally, two sets of triaxial wetting tests on rockfills considering the suction reduction [32] were adopted as comparison to verify the capability to predict the collapse deformation for the proposed model. The tested materials were schists and greywacke, respectively, which were used as the rockfill materials for the inner and outer shell of Beliche Dam. Tests were performed on a dry material first

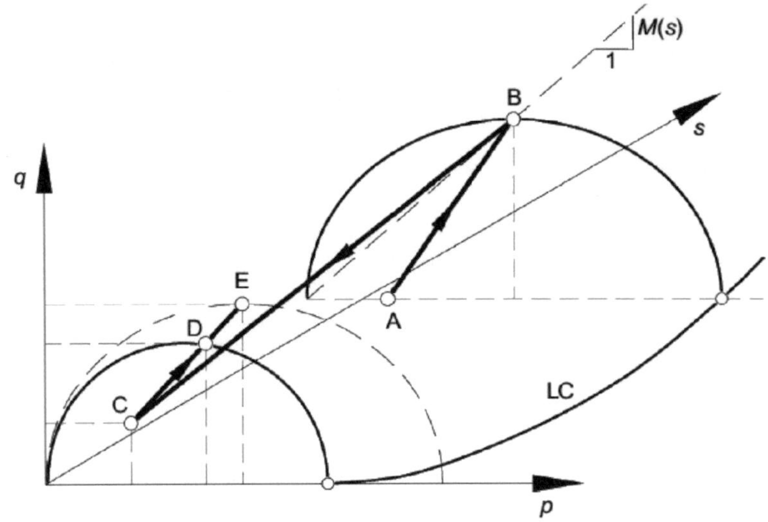

Figure 10.
Sketch showing stress-suction path applied in triaxial tests initially dry and later flooded when limiting conditions are reached.

Elastic parameters	Isotropic compression parameters	Critical state parameters	Default parameters
Inner schists/outer greywacke			
G_0 = 80/100	λ^i-κ = 0.025/0.01 MPa^{-1}	M_0 = 1.75/1.8	m_1 = 0.36/0.4
ν = 0.3/0.3	λ_0^d = 0.028/0.01 MPa^{-1}	M_{res} = 1.2/1.6	m_2 = 0.16/0.2
κ_ψ = 0/0	α_s = 7.0/2.0 ($\times 10^{-3}$ MPa^{-1})	α_M = 0.1/0.5 MPa^{-1}	n = 3.5/3.5
	p_y = 0.01/0.01 MPa	β_ψ = 0.03/0.01	
	p_0^* = 0.3/0.3 MPa	e_{ref}^0 = 0.72/0.87	
	e_0 = 0.538/0.538	p_{ref} = 0.01/0.01 MPa	
		ζ = 0.22/0.15	
		α_e = 0.042/0.01	

Table 3.
Model parameters for inner schists and outer greywacke.

(presumably equilibrated with the relative humidity prevailing in the laboratory, for total suction equal to 20 MPa).

At some stage during the triaxial tests, specimens were then flooded. The applied vertical strain rate was maintained during the sample inundation and subsequent straining. The stress-suction path actually experienced by the specimens in the stress space (p, q and ψ) is illustrated in **Figure 10**. The compacted specimen is initially loaded isotropically under very dry condition (point A). The specimen is

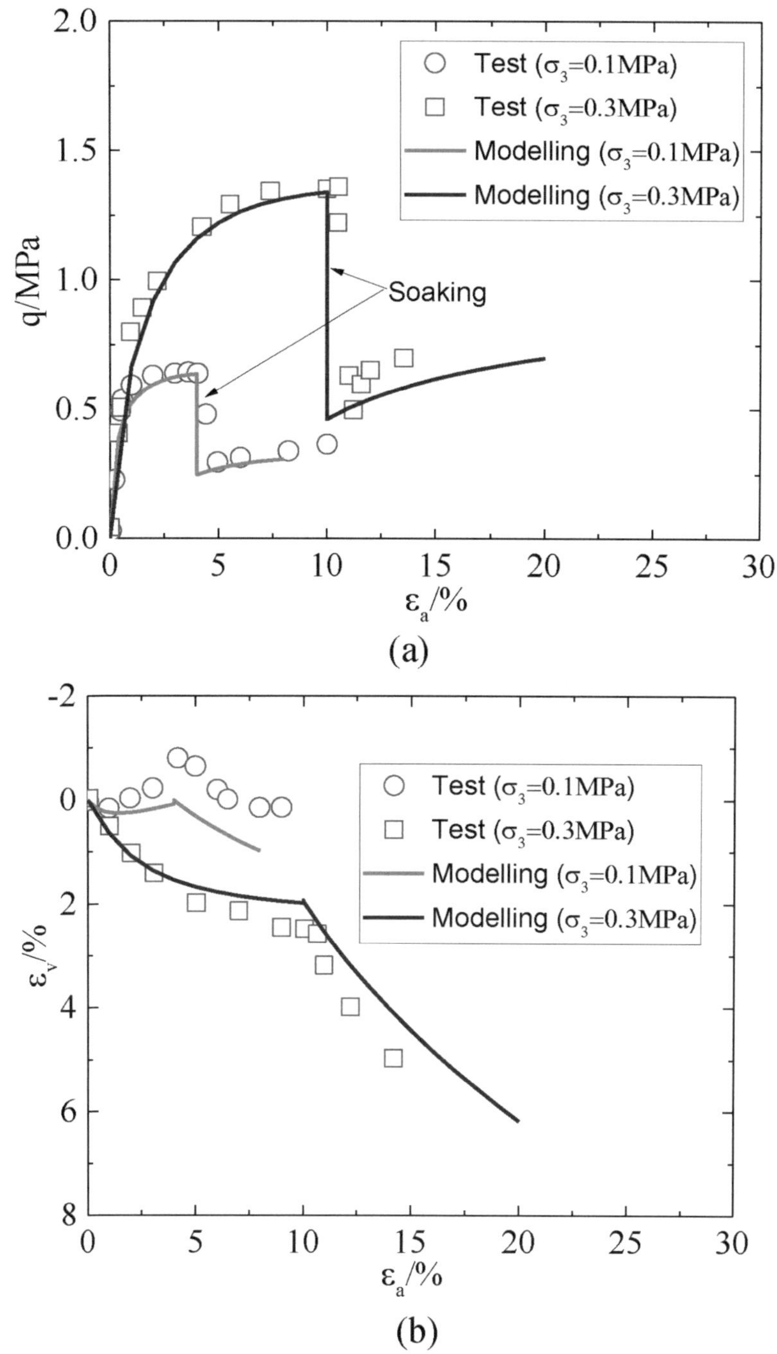

Figure 11.
Triaxial tests on compacted schists from inner shell: (a) q vs. ε_a and (b) ε_v vs. ε_a.

then loaded vertically at constant suction until the limiting condition (point B) is reached. Thereafter, it was flooded with the constant of vertical strain rate maintained. The volumetric collapse induces a sudden loss of vertical stress for the reduction of suction to zero, and Point C is reached. As the total suction reaching zero, the yield envelope experiences a reduction in size. As the vertical strain rate is maintained, the specimen starts resisting again along the loading path and

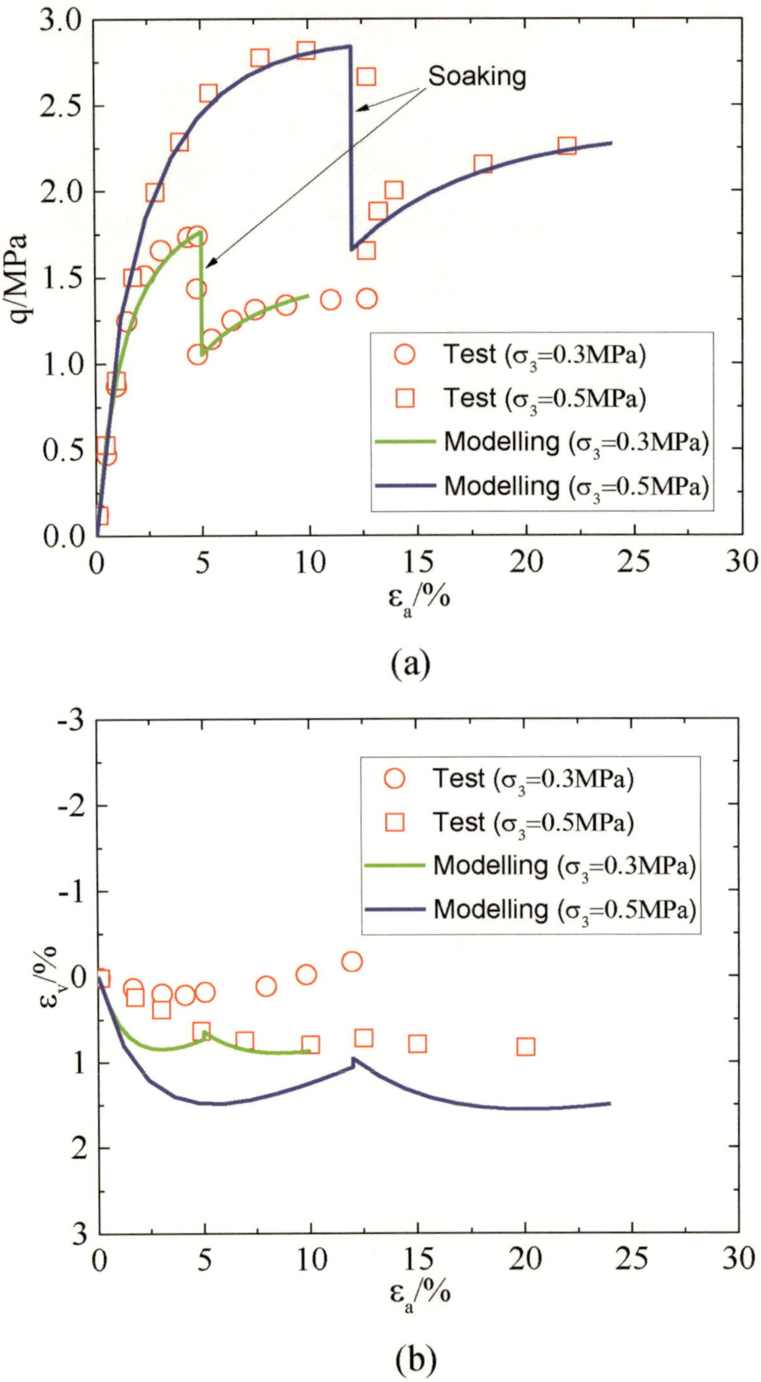

Figure 12.
Triaxial tests on compacted greywacke from outer shell: (a) q *vs.* ε_a *and (b)* ε_v *vs.* ε_a.

eventually reaches a new yield condition at point D, and then a final limiting state at point E. The model parameters for schist and greywacke are listed in **Table 3**, for most of the parameters are referred from Alonso et al. [32], and the rest are determined based on the corresponding experimental results combined with back-analysis method.

The predicted and measured responses of triaxial wetting tests on inner schists and outer greywacke are represented in **Figures 11** and **12**. It can be found that an overall good agreement is achieved between the experimental observations and the model predictions. The measured stress-strain curves in the dry condition, the transient loss of deviatoric stress associated with flooding and the final recovery of strength in the saturated condition can be well captured with the proposed model. However, the dilatancy is underestimated, especially for samples under the low confining pressures. It is worth to note that the shear dilatancy response in the shearing process after flooding for rockfills of outer greywacke can be reproduced by the proposed model, although the predicted compression volumetric strain is a little larger than that measured. In general, the proposed model can reasonably predict the collapse deformation of the unsaturated rockfill materials subjected to the flooding, although it also has a little shortage to predict the volumetric strain.

4. Conclusions

For modelling the strength and deformation behaviour of rockfill materials over various ranges of densities and relative humidity, a state-dependent elastoplastic constitutive model was developed in this study. This model is formulated within an extended critical-state framework by using two independent stress state variables: total stress and total suction. The compressibility model for rockfill materials proposed by Oldecop and Alonso [1] was adopted in this model. The exponential function was used to describe the decrease of the critical stress ratio with the mean stress, while the logarithmic increases of the critical stress ratio and critical void ratio with the total suction were assumed. By substituting these functions into the relation function of peak friction angle with critical friction angle for granular materials proposed by Biarez and Hicher [63], a unified hardening parameter, which could reflect the state-dependent dilatancy, strain-softening and unsaturated effects, was presented for modelling stress-strain relationships of rockfill materials. In addition, the relevant testing procedures have been demonstrated to calibrate and obtain the required model parameters.

The capabilities of the proposed model were illustrated by modelling three series of triaxial tests performed on different rockfill materials (drained triaxial test on saturated samples compacted to various l void ratio, triaxial test with relative humidity controlled and triaxial wetting test). An agreement was obtained between the experimental and numerical results. For the drained triaxial tests on the saturated samples with different initial void ratios, both stress-strain relationship and volumetric strain can be well predicted, which indicates the capability of the proposed model to consider state-dependent behaviour. Through the comparison with the results of relative humidity-controlled triaxial test, the proposed model can be found to well consider the influence of the relative humidity on the stress-strain relationship and volumetric strain of rockfill materials. In addition, the proposed model can well simulate the feature of stress-relaxing according to the comparison with the results of triaxial wetting test. However, the dilatancy is underestimated with the proposed model, especially for the rockfills at the dry state and under the low confining pressures. In general, the model presents an advantage to predict the state-dependent behaviour and collapse deformation of unsaturated rockfills, such as the strain-softening and dilatancy at the dense state. Despite the progress achieved, there is still a room for improvements to be made, for instance, to consider the influence of particle breakage on the changes of compressibility and critical state at different relative humidities.

Author details

Liujiang Wang[1*] and Zhongzhi Fu[2]

1 College of Water Conservancy and Hydropower, Hohai University, China

2 Geotechnical Engineering Department, Nanjing Hydraulic Research Institute, Nanjing, China

*Address all correspondence to: 15850514642@163.com

References

[1] Oldecop LA, Alonso EE. A model for rockfill compressibility. Géotechnique. 2001;**51**(2):127-139

[2] Sowers GF, Williams RC, Wallace TS. Compressibility of broken rock and settlement of rockfills. In: Proceedings of the 6th ICSMFE; Montreal. Vol. 2. 1965. pp. 561-565

[3] Nobari ES, Duncan JM. Effect of reservoir filling on stresses and movements in earth and rockfill dams. Report No. TE-72-1. Department of Civil Engineering, University of California; 1972

[4] Marsal RJ. Mechanical properties of rockfill. In: Hirschfeld RC, Poulos SJ, editors. Embankment Dam Engineering. Casagrande Volume. New York: John Wiley & Sons; 1973. pp. 109-200

[5] Oldecop LA, Alonso E(AP d A). Suction effects on rockfill compressibility. Géotechnique. 2003; **53**(2):213-234

[6] Oldecop LA, Alonso EE. Testing rockfill under relative humidity control. Geotechnical Testing Journal. 2004; **27**(3):269-278

[7] Chávez C, Alonso EE. A constitutive model for crushed granular aggregates which includes suction effects. Soils and Foundations. 2003;**43**(4):215-227

[8] Chávez C, Romero E, Alonso EE. A rockfill triaxial cell with suction control. Geotechnical Testing Journal. 2009; **32**(3):219-231

[9] Bauer E, Fu ZZ, Liu SH. Hypoplastic constitutive modelling of wetting deformation of weathered rockfill materials. Frontiers of Architecture and Civil Engineering in China. 2010;**4**(1): 78-91

[10] Kohgo Y, Asano I, Hayashida Y. Mechanical properties of unsaturated low quality rockfills. Soils and Foundations. 2007a;**47**(5):947-959

[11] Marsal RJ. Large scale testing of rockfill materials. Journal of the Soil Mechanics and Foundations Division. 1967;**93**(2):27-43

[12] Charles JA, Watts KS. The influence of confining pressure on the shear strength of compacted rockfill. Géotechnique. 1980;**30**(4):353-367

[13] Dine BSE, Dupla JC, Frank R, et al. Mechanical characterization of matrix coarse-grained soils with a large-sized triaxial device. Canadian Geotechnical Journal. 2010;**47**(4):425-438

[14] Weng MC, Chu BL, Ho YL. Elastoplastic deformation characteristics of gravelly soils. Journal of Geotechnical and Geoenvironmental Engineering. 2013;**139**(6):947-955

[15] Xiao Y, Liu H, Chen Y, et al. Bounding surface model for rockfill materials dependent on density and pressure under triaxial stress conditions. Journal of Engineering Mechanics. 2014a;**140**(4):04014002

[16] Alonso EE, Romero EE, Ortega E. Yielding of rockfill in relative humidity-controlled triaxial experiments. Acta Geotechnica. 2016;**11**(3):455-477

[17] Varadarajan A, Sharma KG, Abbas SM, et al. Constitutive model for rockfill materials and determination of material constants. International Journal of Geomechanics. 2006;**6**(4):226-237

[18] Honkanadavar NP, Sharma KG. Testing and modeling the behavior of riverbed and blasted quarried rockfill materials. International Journal of Geomechanics. 2014;**14**(6):04014028

[19] Fu Z, Chen S, Peng C. Modeling cyclic behavior of rockfill materials in a

framework of generalized plasticity. International Journal of Geomechanics. 2014;**14**(2):191-204. DOI: 10.1061/(ASCE)GM.1943-5622.0000302

[20] Tosun H. Discussion of "construction of concrete-faced rockfill dams with weak rocks" by Hao-Feng Xing, Xiao-Nan Gong, Xiao-Guang Zhou, and Hai-Feng Fu. Journal of Geotechnical and Geoenvironmental Engineering. 2006;**132**(6):778-785. DOI: 10.1061/ASCE1090-02412006132:6778

[21] Naylor DJ, Tong SL, Shahkarami AA. Numerical modelling of saturation shrinkage. In: Proceedings of 3rd International Symposium on Numerical Models in Geomechanics; Niagara Falls. 1989. pp. 636-648

[22] Shen ZJ, Wang JP. Numerical simulation of construction behaviour of clay core dam and its movement due to reservoir impounding. Hydro-Science and Engineering. 1988;**4**(1):47-64

[23] Li GY, Wang LS, Mi ZK. Research on stress-strain behaviour of soil core rockfill dam. Chinese Journal of Rock Mechanics and Engineering. 2004; **23**(8):1363-1369

[24] Fang XS. Test study and numerical simulation on wetting deformation of gravel sand [dissertation of masteral degree]. Nanjing: Hohai University; 2005

[25] Wei S. Study on wetting deformation behaviour and numerical model of coarse-grained materials [dissertation of doctoral degree]. Nanjing: Hohai University; 2006

[26] Naylor DJ, Maranha JR, Neves EMD, et al. A back-analysis of Beliche dam. Géotechnique. 2001;**51**(4): 377-381

[27] Fu ZZ, Chen SS, Liu SH. Hypoplastic constitutive modelling of the wetting induced creep of rockfill materials. Science China Technological Sciences. 2012;**55**(7):2066-2082

[28] Marachi ND, Chan CK, Seed HB, et al. Strength and deformation characteristics of rockfill materials. Report No. TE-69-5. Department of Civil Engineering, University of California; 1969

[29] Terzaghi K. Discussion on settlement of salt springs and Lower Bear River concrete face dams. Transactions of the American Society of Civil Engineers. 1960;**123**(2):139-148

[30] Alonso EE. A constitutive model for partially saturated soils. Géotechnique. 1990;**40**(3):405-430

[31] Alonso EE, Olivella S, Pinyol NM. A review of Beliche Dam. Géotechnique. 2005;**55**(4):267-285

[32] Alonso EE, Olivella S, Pinyol NM, et al. Modelling the response of Lechago earth and rockfill dam. Géotechnique. 2011;**61**(5):387-407

[33] Kohgo Y, Asano I, Hayashida Y. An elastoplastic model for unsaturated rockfills and its simulations of laboratory tests. Soils and Foundations. 2007;**47**(5):919-929

[34] Bauer E. Hypoplastic modelling of moisture-sensitive weathered rockfill materials. Acta Geotechnica. 2009;**4**(4): 261-272

[35] Jefferies MG. Nor-sand: A simple critical state model for sand. Géotechnique. 1993;**43**(1):91-103

[36] Bauer E. Calibration of a comprehensive hypoplastic model for granular materials. Soils and Foundations. 1996;**36**(1):13-26

[37] Gudehus G. A comprehensive constitutive equation for granular materials. Journal of the Japanese Geotechnical Society. 1996;**36**(1):1-12

[38] Wu W, Bauer E, Kolymbas D. Hypoplastic constitutive model with critical state for granular materials. Mechanics of Materials. 1996;**23**(1): 45-69

[39] Manzari MT, Dafalias YF. A critical state two-surface plasticity model for sands. Géotechnique. 1997;**47**(2): 255-272

[40] Gajo A, Wood DM. Severn-Trent sand: A kinematic-hardening constitutive model: The q-p formulation. Géotechnique. 1999;**49**(5): 595-614

[41] Li XS, Dafalias YF. Dilatancy for cohesionless soils. Géotechnique. 2000; **50**(4):449-460

[42] Wang ZL, Dafalias YF, Li XS, et al. State pressure index for modeling sand behavior. Journal of Geotechnical & Geoenvironmental Engineering. 2002; **128**(6):511-519

[43] Yao YP, Sun DA, Luo T. A critical state model for sands dependent on stress and density. 2004;**28**(4):323-337

[44] Loukidis D, Salgado R. Modeling sand response using two-surface plasticity. Computers and Geotechnics. 2009;**36**(1–2):166-186

[45] Chakraborty T, Salgado R, Loukidis D. A two-surface plasticity model for clay. Computers and Geotechnics. 2013;**49**(APR.):170-190

[46] Xiao Y, Liu H, Chen Y, et al. Testing and modeling of the state-dependent behaviors of rockfill material. Computers and Geotechnics. 2014; **61**(1):153-165

[47] Sun Y, Gao Y, Zhu Q. Fractional order plasticity modelling of state-dependent behaviour of granular soils without using plastic potential. International Journal of Plasticity. 2017; (3):53-69

[48] Liu SH, Wang LJ, Wang ZJ, Bauer E. Numerical stress-deformation analysis of cut-off wall in clay-core rockfill dam on thick overburden. Water Science and Engineering. 2016;**9**(3):219-226

[49] Wang LJ, Liu SH, Fu ZZ, Li Z. Coupled hydro-mechanical analysis of slope under rainfall using modified elasto-plastic model for unsaturated soils. Journal of Central South University. 2015;**22**(5):1892-1900

[50] Wang L, Liu S, Liao J, et al. Field load tests and modelling of soft foundation reinforced by soilbags. Geosynthetics International. 2019:1-36

[51] Sheng D, Sloan SW, Gens A, et al. Finite element formulation and algorithms for unsaturated soils. Part 1: Theory. International Journal for Numerical and Analytical Methods in Geomechanics. 2003a;**27**(9):745-765

[52] Coussy O. Mechanics of Porous Continua. Chichester, UK: Wiley; 1995

[53] McDowell GR, Bolton MD. On the micromechanics of crushable aggregates. Géotechnique. 1998;**48**(5): 667-679

[54] Yao YP, Hou W, Zhou AN. UH model: Three-dimensional unified hardening model for overconsolidated clays. Géotechnique. 2009;**59**(5): 451-469

[55] Indraratna B. Large-scale triaxial testing of greywacke rockfill. Géotechnique. 1993;**43**(1):37-51

[56] Frossard E, Dano C, Hu W, et al. Rockfill shear strength evaluation: A rational method based on size effects. Géotechnique. 2012;**62**(5):415-427

[57] Xiao Y, Liu HL, Zhu JG, et al. Modeling and behaviours of rockfill materials in three-dimensional stress space. Science China Technological Sciences. 2012;**55**(10):211-226

[58] Yao YP, Sun DA, Matsuoka H. A unified constitutive model for both clay and sand with hardening parameter independent on stress path. Computers and Geotechnics. 2008;**35**(2):210-222

[59] Yao YP, Zhou AN. Non-isothermal unified hardening model: A thermo-elasto-plastic model for clays. Géotechnique. 2013;**63**(15):1328-1345

[60] Biarez J, Hicher P-Y. Elementary Mechanics of Soil Behaviour. Rotterdam, Netherlands: A.A. Balkema; 1994

[61] Hicher PY, Chang CS. A microstructural elastoplastic model for unsaturated granular materials. International Journal of Solids and Structures. 2007;**44**(7–8):2304-2323

[62] Sheng D, Smith DW, Sloan SW, et al. Finite element formulation and algorithms for unsaturated soils. Part II: Verification and application. International Journal for Numerical and Analytical Methods in Geomechanics. 2003b;**27**(9):767-790

[63] Naylor DJ, Maranha DNE, Veiga Pinto AA, et al. Prediction of construction performance of Beliche dam. Géotechnique. 1986;**36**(3):359-376

Chapter 6

A Practical Framework for Probabilistic Analysis of Embankment Dams

Xiangfeng Guo and Daniel Dias

Abstract

Uncertainties, such as soil parameters variability, are often encountered in embankment dams. Probabilistic analyses can rationally account for these uncertainties and further provide complementary information (e.g., failure probability and mean/variance of a model response) than deterministic analyses. This chapter introduces a practical framework, based on surrogate modeling, for efficiently performing probabilistic analyses. An active learning process is used in the surrogate model construction. Two assessment stages are included in this framework by respectively using random variables (RV) and random fields (RF) for the soil variability modeling. In the first stage, a surrogate model is coupled with three probabilistic methods in the RV context for the purpose of providing a variety of useful results with an acceptable computational effort. Then, the soil spatial variability is considered by introducing RFs in the second stage that enables a further verification on the structure reliability. The introduced framework is applied to an embankment dam stability problem. The obtained results are validated by a comparison with direct Monte Carlo Simulations, which also allows to highlight the efficiency of the employed methods.

Keywords: embankment dam, slope stability, reliability analysis, sensitivity analysis, random field

1. Introduction

According to the International Commission of Large Dams (ICOLD) database updated in September 2019 [1], there are around 58,000 large dams (higher than 15 m) over the world and 75% of them can be classified as embankment dams. Concerning all the constructed dams, the number is much more important. For example, over 91,460 dams were operated across the United States in 2019 [2] and the majority is rock-filled or earth-filled ones. Therefore, safety assessment of embankment dams is crucial for engineers considering their great population and the considerable damages that can be induced by their failures. However, embankment dams involve a high degree of uncertainties, especially for their material properties [3] since they are constructed by natural materials (soils, sands, or rocks), which makes their safety evaluation a difficult task. Probabilistic analysis [4] is an effective solution which permits to rationally account for the soil variabilities and quantify their effects on the dam safety condition by using a reliability method or a sensitivity method. Additionally, complementary results [5] can be provided by a probabilistic analysis compared to a traditional deterministic assessment, including the failure probability (*Pf*), design point, model response statistics (e.g., mean and variance) and sensitivity index. Having more results are beneficial for designers to better understand the functioning mode of the dam and make more rational decisions. Therefore, it is worthy to implement probabilistic analyses for the safety assessment of embankment dams, in order to account for the soil variabilities and provide complementary information. **Figure 1** shows a comparison between a probabilistic and a deterministic analysis. In this figure, MCS and FORM [4] are two reliability methods respectively referring to Monte Carlo Simulation and First Order Reliability Method. FoS represents the factor of safety and can be replaced by other types of model responses (like the settlement) which are also of interest for engineers.

In a probabilistic analysis, uncertainties of soil properties can be represented by random variables (RVs) or random fields (RFs) [4]. The former is simpler and easier to couple with a deterministic model [4]. In the RV approach, the soil is assumed to be homogeneous but different values are generated in different simulations for one soil property according to a given distribution. Therefore, the RV method cannot explicitly account for the soil spatial variabilities. On the contrary, the RF approach can model the spatial variation of soils. For a soil property in one simulation, one RF, meaning a collection of different values in a discretized grid, is generated according to the soil parameter statistics and a given autocorrelation structure. However, this approach is more complex and needs extra computational efforts (e.g., quantification of the autocorrelation distances and generation of RFs) compared to the RV one. **Figure 2** illustrates the principle idea of the two approaches.

In this chapter, a practical framework is proposed to efficiently perform the probabilistic analysis of embankment dams. The RV and RF approaches are both implemented into the framework, corresponding to two assessment stages. The RV approach permits a quick estimate on the target results (e.g., *Pf*) while the RF one is able to account for the soil spatial variability and update the *Pf* in order to be more precise in a second stage. The proposed framework is applied to an embankment dam stability problem to show its capacity of providing many useful results and its high computational efficiency. A discussion section is provided as well in which the obtained results are validated by comparing with direct MCS. Besides, some issues such as the reliability method selection and probabilistic analysis tools are discussed.

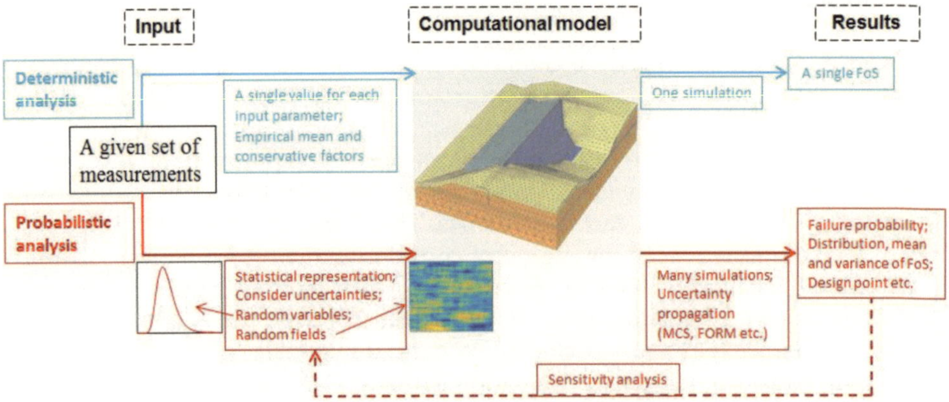

Figure 1.
Contributions of a probabilistic analysis (red color) to a deterministic analysis (blue color).

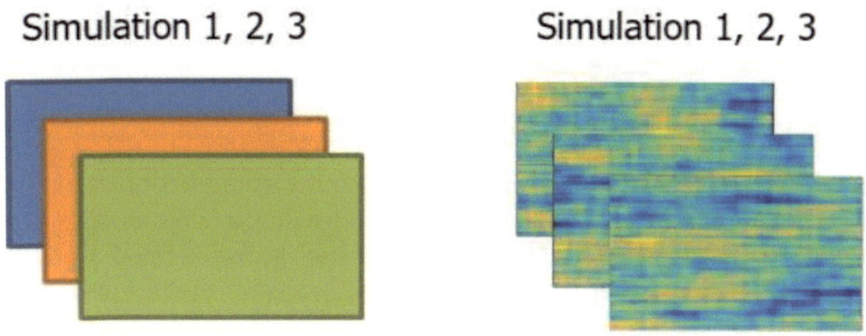

Figure 2.
Comparison between the two approaches of uncertainty modeling (left: Random variables; right: Random fields).

2. Presentation of the used probabilistic analysis methods

This section aims to briefly present the probabilistic analysis methods used in the proposed framework including two reliability methods (MCS and FORM), a surrogate modeling technique (PCE), a global sensitivity analysis method (Sobol) and a RF generation approach (KLE).

2.1 Monte Carlo simulations (MCS)

The MCS offers a robust and simple way to estimate the distribution of a random model response and assess the associated Pf. The idea is to largely and randomly generate samples according to a joint input Probability Density Function (PDF) and evaluate the model response of each sample (i.e., an input vector $\boldsymbol{x}$) by a deterministic computational model. For an MCS with N_{MC} model evaluations, the Pf can be approximated by [4]:

$$Pf \approx \frac{1}{N_{MC}} \cdot \sum_{i=1}^{N_{MC}} I_{MC}\left(\boldsymbol{x}^{(i)}\right) \tag{1}$$

where $I_{MC}(\cdot)$ is an indicator function with $I_{MC}(x) = 1$ if x leads to a failure, otherwise $I_{MC}(x) = 0$. The value of N_{MC} should be large enough in order to obtain an accurate estimate for the Pf which can be assessed by its Coefficient of Variation (CoV):

$$CoV_{Pf} = \sqrt{(1 - Pf)/(N_{MC} \cdot Pf)} \cdot 100\% \tag{2}$$

It is important to mention that the CoV_{Pf} of Eq. (2) is independent of the problem dimension. Additionally, the MCS works regardless of the complexity of the Limit State Surface (LSS). However, a crude MCS suffers from a low computational efficiency. According to Eq. (2), around $100/Pf$ model evaluations are required if the target CoV_{Pf} is 10%.

2.2 First order reliability method (FORM)

The FORM estimates the Pf by approximating the LSS locally at a reference point with a linear expansion. The reference point is called as design point $\boldsymbol{P}^*$. It is defined in the standard normal space as the point that is on the LSS and closest to the space origin O_{SN}. This point can be located by solving an optimization problem as [6]:

$$P^* = \text{argmin}\{\|u\|, G(u) \le 0\} \tag{3}$$

where u is the input vector x transformed into the standard normal space and $G(\cdot)$ is the performance function with $G(u) \le 0$ representing the failure domain. For the slope stability analysis, the performance function can be defined as: $G = FoS - 1$. Once the P^* is determined, the Pf can be approximated by the following equation:

$$Pf \approx \Phi_{SN}(-\beta_{HL}) = \Phi_{SN}(-\|P^*\|) \tag{4}$$

where Φ_{SN} is the standard normal Cumulative Density Function (CDF) and β_{HL} is the Hasofer-Lind reliability index. Additionally, based on the components of the vector from O_{SN} to P^*, the importance factor of each RV can be derived [6].

2.3 Polynomial Chaos expansions (PCE)

The PCE is a powerful and efficient tool for metamodeling which consists in building a surrogate of a complex computational model. It approximates a model response Y by finding a suitable basis of multivariate orthonormal polynomials with respect to the joint input PDF in the Hilbert space. The basic formula of PCE is [7]:

$$Y \approx \sum_{\alpha \in \mathbb{N}^M} \boldsymbol{k_\alpha} \Psi_\alpha(\boldsymbol{\xi}) \tag{5}$$

where $\boldsymbol{\xi}$ are independent RVs, $\boldsymbol{k_\alpha}$ are unknown coefficients to be computed with α being a multidimensional index and Ψ_α are multivariate polynomials which are the tensor product of univariate orthonormal polynomials. The representation of Eq. (5) should be truncated to a finite number of terms for practical applications by using the standard or hyperbolic truncation scheme. Then, the unknown coefficients can be estimated by using the Least Angle Regression method. The accuracy of the truncated PCE can be assessed by computing the coefficient of determination R^2 and the Q^2 indicator: R^2 is related to the empirical error using the model responses already existing in the design of experiment (DoE), while Q^2 is obtained by the leave-one-out cross-validation technique [7].

In order to further reduce the number of Ψ_α after the truncate operation when the input dimension is high, the sparse PCE (SPCE) was proposed [7]. The idea came from the fact that the non-zero coefficients in the PCE form a sparse subset of the truncation set obtained by the hyperbolic truncation scheme. Thus, it consists in building a suitable sparse basis instead of computing useless terms in the expansions that are eventually negligible.

2.4 Sobol-based global sensitivity analysis (GSA)

The GSA aims to evaluate the sensitivity of a Quantity of Interest (QoI) with respect to each RV over its entire varying range. Among many methods for performing a GSA, the Sobol index has received much attention since they can give accurate results for most models [8]. The Sobol-based GSA is based on the variance decomposition of the model output . The first order Sobol index is given as:

$$S_i = \frac{V_i(Y)}{V_t} = \frac{Var[\mathbb{E}(Y\|x_i)]}{V_t} \tag{6}$$

where V_t is the total variance of Y. For the $Var[\mathbb{E}(Y\|x_i)]$, the inner expectation operator $\mathbb{E}(\cdot)$ is the mean of Y considering all possible $\boldsymbol{x}_{\sim i}$ values while keeping x_i constant; the outer variance $Var(\cdot)$ is taken over all possible values of x_i. The first order Sobol index measures the contribution of the variable x_i solely. Another important parameter in a Sobol-based GSA is the total effect index which is given as:

$$S_{Ti} = S_i + \sum_{i \neq j} S_{ij} + \dots S_{1,\dots,M} \tag{7}$$

where S_{ij}, ... $S_{1,\dots,M}$ represents the higher order Sobol index. S_{Ti} is able to take into account the interaction effects of the variable x_i with other variables.

It is noted that the Sobol index is only effective for independent variables. In order to properly account for the input correlation effect, the Kucherenko index [5] can be employed which is also based on the variance decomposition. For the estimation of the Sobol or Kucherenko index (First order and total effect), the traditional way is to use the idea of MCS however it requires a high number of model evaluations.

2.5 Karhunen-Loève expansions (KLE)

A random field (RF) can describe the spatial correlation of a material property in different locations and represent nonhomogeneous characteristics. The KLE, as a series expansions method, is widely used in the geotechnical reliability analysis since it can lead to the minimal number of RVs involved in a RF discretization [7]. In the KLE context, a stationary Gaussian RF H can be expressed as follows:

$$H(\boldsymbol{x_{RF}}) \approx \mu + \sigma \sum_{i=1}^{N_{KL}} \sqrt{\lambda_i} \phi_i(\boldsymbol{x_{RF}}) \xi_i \tag{8}$$

where x_{RF} is the coordinate of an arbitrary point in the field, μ and σ represents respectively the mean and standard deviation of the RF, λ_i and ϕ_i are respectively the eigenvalues and eigenfunctions of the autocovariance function for the RF, ξ_i is a set of uncorrelated standard normal RVs and N_{KL} is used to truncate the KLE for practical applications. The autocovariance function is the autocorrelation function multiplied by the RF variance. The 2D exponential autocorrelation function is commonly used in the field of reliability analysis. It can be given by:

$$\rho(\boldsymbol{x}, \boldsymbol{x'}) = \exp\left(-\frac{|x - x'|}{L_x} - \frac{|y - y'|}{L_z}\right) \tag{9}$$

where (x, y) and (x', y') are the coordinates of two arbitrary points in the RF, L_x and L_z is respectively the horizontal and vertical autocorrelation distance. The autocorrelation distance is defined as the length which can lead to a decrease from 1 to 1/e for the autocorrelation function. Concerning the N_{KL}, its value is determined by evaluating the error due to the truncation term. The variance-based error globally estimated in the RF domain Ω can be expressed as [9]:

$$\varepsilon_{KL} = \frac{1}{\Omega} \int_{\Omega} \left[1 - \sum_{i=1}^{N_{KL}} \lambda_i \phi_i^2(\boldsymbol{x_{RF}}) \right] d\Omega \tag{10}$$

3. The introduced framework

This section presents the introduced framework for the probabilistic analysis of embankment dams. A flowchart of the framework is given in **Figure 3**.

At the beginning, three elements should be prepared. Firstly, the distribution type and the related parameters (e.g., mean and variance) of the concerned material properties have to be determined. It will allow describing their uncertainties by means of RVs. The selected material properties should be relevant to the QoI of the problem. In case of it is difficult to properly select the relevant properties, all the possible properties can be considered for the RV modeling. The Global Sensitivity Analysis (GSA) which will be performed in the first stage can help to understand the significance of each property. With the GSA results, one can then select which properties will be modeled by RFs. The second work is to develop a deterministic computational model by using numerical or analytical methods (e.g., Finite element method and Limit analysis method). The objective of this model is to estimate the QoI with a given set of input parameters. Then, the autocorrelation structure of the concerned properties should be determined. This structure, defined by an autocorrelation function and the autocorrelation distances, allows to describe the spatial correlation between different locations of a property. It is a key element in the generation of RFs. After these three preparation-works, the analyses in the two stages can be performed. It should be noted that the focus of this chapter is to show the benefits of a probabilistic analysis and demonstrate the proposed framework. Concerning the way of rationally determining the distribution parameters and the autocorrelation structure by using the available measurements, readers can refer to [10, 11].

The objective of the first stage is to provide a variety of probabilistic results with an acceptable computational burden. The results could be helpful to analyze the current problem in a preliminary design phase and guide the following site investigation program and the next design assessment phase. In this stage, the RV approach is used to consider the input uncertainties. It allows quickly having a first

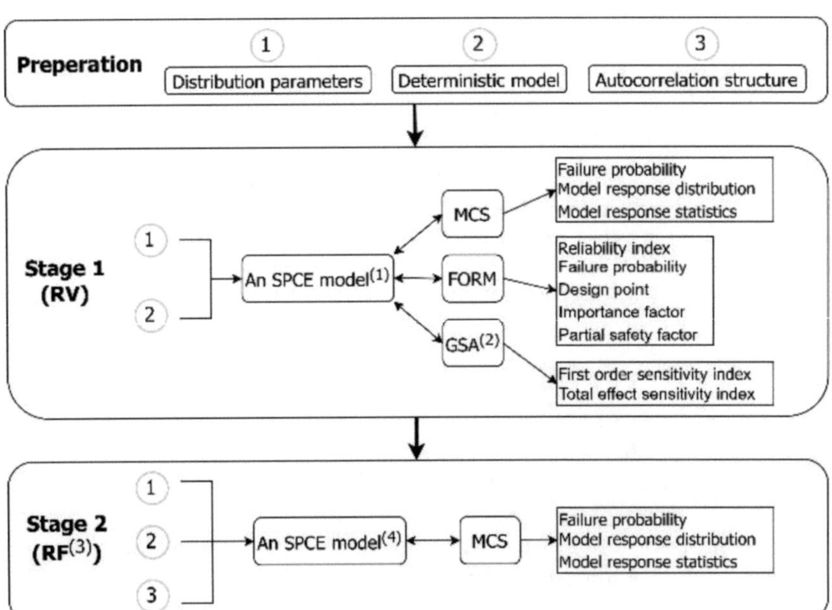

Figure 3.
Flow chart of the proposed framework.

view on the target results given that this approach can be easily coupled with any deterministic model and any probabilistic analysis method. Three analyses are performed in this stage by using respectively three techniques: two reliability methods (MCS and FORM) and one sensitivity method (Sobol-based GSA). The MCS is always considered as a reference method to evaluate other reliability methods due to its robustness. Therefore, an MCS is conducted here in order to obtain an accurate estimate on the Pf. It can also provide the model response distribution and statistics. The FORM is an approximation method due to its linear assumption. It is also adopted in this stage because this method can provide a variety of valuable results which could be beneficial for engineers. For example, the design point permits to know how much margin there are with respect to the current mean values, and the partial safety factors are comparable with the conservative factors used in a deterministic analysis to penalize the strength properties. The Sobol-based GSA permits to quantify the contribution of each RV to the model response variance. By using the Kucherenko index, the correlation effect among the RVs can also be accounted for. According to the GSA results, the properties, which have very slight effects, could be kept as RVs or treated as deterministic in the second stage. This can significantly reduce the computational burden. Particularly, the three analyses are conducted by using a surrogate model (SPCE). The aim is to reduce the total computational time given that a direct MCS or GSA is very time-consuming since they need usually tens of thousands of deterministic model evaluations. For most cases, it is not affordable to repeatedly run a deterministic model with a number higher than 10^4. Besides, an active learning process [12] given in **Figure 4**, is used to construct the SPCE model. This process starts with an initial DoE and gradually enriches it by adding new samples. A new SPCE model is created each time after the DoE updated with new samples. This process is stopped when some criteria are satisfied. This algorithm is more efficient than the metamodel training based on a single DoE and can give accurate estimate on the Pf.

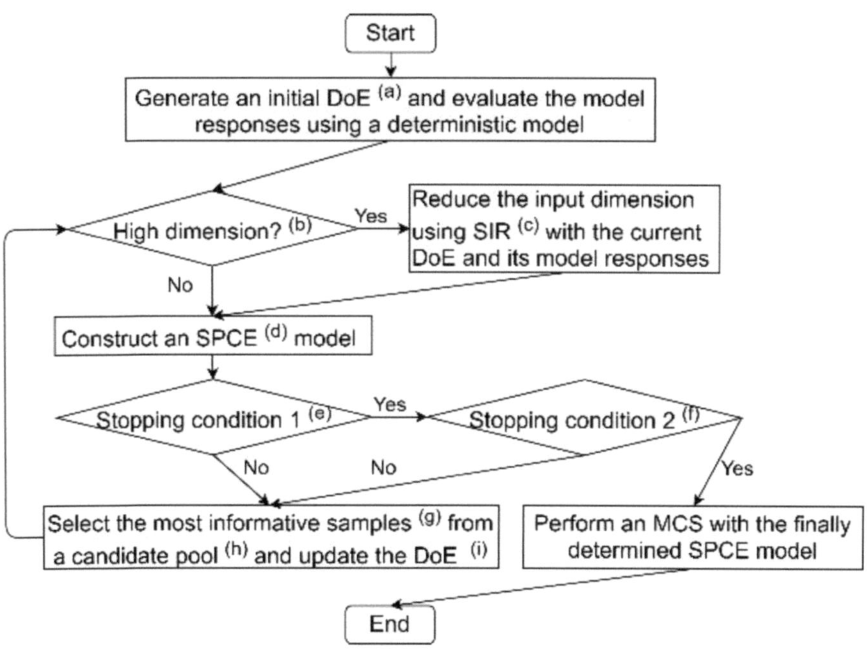

Figure 4.
An active learning process for the PCE model construction.

Figure 3	1.	The procedure of **Figure 4** is followed to create an SPCE model
	2.	For independent RVs, the Sobol index is used For correlated RVs, the Kucherenko index is used
	3.	The RFs are generated by the KLE Conditional RFs can be used if knowing the measurements' locations
	4.	The SIR is used to reduce the input dimension An SPCE is created in the reduced space
Figure 4	a.	The size N_{ini} of the initial DoE is: max [12, M] M: input dimension Latin Hypercube Sampling (LHS) is used for generating samples
	b.	There is no standard value to check if the problem is high dimensional or not. For the SPCE model, it can easily handle 10–20 RVs. It is thus better to reduce the dimension if $M > 20$ or 30.
	c.	An important parameter in the SIR is the slice number N_{sir} $10 \leq \text{Nsir} \leq 20$ for the cases with several hundred RVs [9] $20 \leq \text{Nsir} \leq 30$ when the number of input RVs is several thousands
	d.	The algorithm presented in [9] is used to create an SPCE The SPCE optimal order is determined by testing in a range
	e.	Stopping condition 1 measures if the accuracy indicator Q^2 of the constructed SPCE model is higher than a target value Q_t^2.
	f.	Stopping condition 2 evaluates the convergence of the Pf estimation by computing an error Err_{con} which is the maximum value of the relative errors calculated from all the possible pairs in a vector. The vector consists of the N_{s2} last Pf estimates in the adaptive DoE process. The condition will be satisfied if Err_{con} is lower than a given value Err_t.
	g.	N_{add} samples are selected by using the strategy of [12].
	h.	An MCS population is generated using the LHS as a candidate pool
	i.	DoE is updated by adding the new samples and their model responses

Table 1.
Details remarks to some specific features of ***Figures 3*** *and* ***4****.*

The second stage aims to consider the spatial variation of the concerned properties which are ignored in the previous stage. It can thus provide a more precise Pf estimate in a second (final) design phase. The new data collected in the new site investigation can be incorporated in this stage in order to update the uncertainty modeling. The GSA results of the first stage can be used to reduce the number of the properties that should be modeled by RFs. The probabilistic analysis becomes a high dimensional problem in this stage due to the RF discretization. As a result, only the MCS is used since the other two methods have difficulties of handling a large number of input RVs (high dimension). The SPCE coupled with the adaptive DoE process is also used at this stage in order to accelerate the MCS. Particularly, a dimension reduction technique – Sliced Inverse Regression (SIR) [9] is used to reduce the input dimension. The SIR is based on the principle that a few linear combinations of original input variables could capture the essential information of model responses. **Table 1** gives a summary of the specific remarks to **Figures 3** and **4**.

4. Application to an embankment dam example

This section shows an application of the proposed framework to an embankment dam stability problem. The dam initially proposed and studied in [5, 13] is selected for this application.

4.1 Presentation of the studied dam and deterministic model

The studied dam is given in **Figure 5**. It has a width of 10 m for the crest and a horizontal filter drain installed at the toe of the downstream slope. The soil is assumed to follow a linear elastic perfectly plastic behavior characterized by the Mohr Coulomb shear failure criterion. In this work, the dam stability issue will be analyzed by considering a constant water level of 11.88 m and a saturated flow. Additionally, a horizontal pseudo-static acceleration of 2.16 m/s^2 toward the downstream part is applied on the dam body. This value represents a relatively high seismic loading and is determined by referring to the recommendations given in [14] for a dam of category A with a soil of type B.

Concerning the input uncertainty modeling, three soil properties (density γ, effective cohesion c' and friction angle ϕ') of the compacted fill are modeled by lognormal RVs or RFs. The illustrative values for the distribution parameters and autocorrelation distances are given in **Table 2**. The uncertainties in the soil hydraulic parameters are not considered since the variation of the dam phreatic level in the downstream part is not significant due to the presence of the filter drain. The mean values are taken from a real dam case reported in [10] and the selected CoVs are consistent with the recommendations give in [3]. A correlation coefficient of -0.3 is considered between c' and ϕ' since it usually exists a negative correlation between these two properties and the correlation coefficient is varied with a range of $[-0.2, -0.7]$ [5]. The L_x is assumed to be significantly larger than L_z since embankment dams are constructed by layers and the spatial variation of material properties is less remarkable in the horizontal direction than the vertical one. The other soil properties are considered as deterministic by using the values given in [5].

The deterministic model used in this work for estimating the dam FoS is developed by using the idea of [13]. It combines three techniques: Morgenstern Price Method (MPM), Genetic Algorithm (GA) and a non-circular slip surface generation

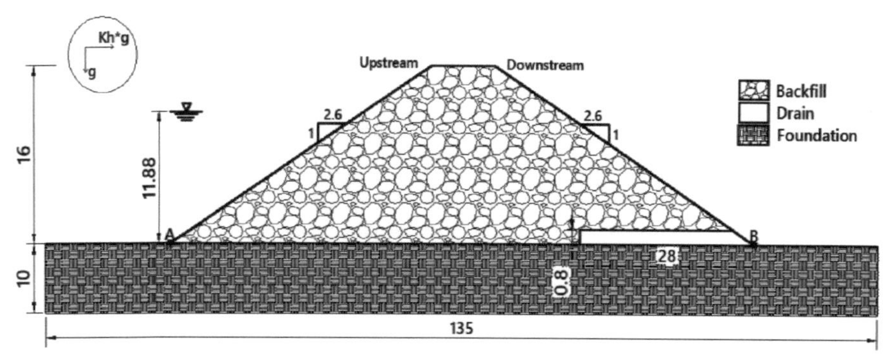

Figure 5.
Geometry of the studied dam (g: gravity acceleration).

Soil property	Distribution	Mean	CoV	Correlation coefficient	L_x (m)	L_z (m)
γ (kN/m^3)	Lognormal	19.8	5%		40	8
c' (kPa)	Lognormal	8.9	30%	-0.3	40	8
ϕ' (degree)	Lognormal	34.8	10%		40	8

Table 2.
Distribution parameters and autocorrelation distance for γ, c' and ϕ'.

method. MPM is employed to compute the FoS of a given failure surface; GA aims at locating the most critical slip surface (i.e., minimum FoS) by performing an optimization work; The implementation of non-circular slip surfaces can lead to more rational failure mechanics for the cases of non-homogeneous soils. The principle of the model is to firstly generate a number of trial slip surfaces as an initial population, and then to determine the minimum FoS value by modeling a natural process along generations including reproduction, crossover, mutation and survivors' selection. The distribution of the pore water pressures inside the dam is given by a numerical model [5]. In this work, the developed deterministic model is termed as LEM-GA. Using a simplified deterministic model (e.g., LEM-GA) is beneficial for a reliability analysis since it can reduce the total computational time. This strategy can thus be adopted in a preliminary design/assessment phase for efficiently obtaining first results. Then, a sophisticated model (e.g., Finite element model) is required in a next phase if complex conditions should be modeled (e.g., rapid drawdown and unsaturated flows) or multiple model responses (e.g., settlement and flow rate) are necessary.

4.2 First stage: RV approach

This section shows the conducted works at the first stage of the proposed framework and presents the obtained results. The RV approach is used in order to have a quick estimate on the dam reliability and the contribution of each input variable. The joint input PDF is defined by the mean, CoV and $\beta_{c\phi}$ of the three soil properties given in **Table 2**. Three probabilistic analyses (MCS, FORM, and GSA) are performed with a surrogate model, also known as meta-model, in this stage so that a variety of useful results can be obtained.

Firstly, an SPCE surrogate model is constructed as an approximation to the model LEM-GA. It is achieved by using the procedure of **Figure 4** with the following user-defined parameters: $Q_t^2 = 0.98$; $Err_t = 0.1$; N_{s2}=10; $N_{add} = 1$; $N_{ini} = 12$; the size for the MCS candidate pool is large enough so that the estimated Pf has a CoV lower than 5%. The finally obtained SPCE is a 3-order model with a Q^2 of 0.99. Twelve new samples, determined by the active learning process, are added to the initial DoE which corresponds to a total number of model evaluation (N_{me}) of 24. Then, the SPCE model is respectively coupled with MCS, FORM and GSA in order to provide different results. As a meta-model is usually expressed analytically, the SPCE-based analyses are thus very fast. Therefore, the main computation burden of the first stage lies in constructing a satisfactory SPCE model. In this work, only 24 deterministic calculations are performed for the construction, representing a significant reduction in N_{me} compared to direct MCS, FORM and GSA which require at least tens thousands of model evaluations. This shows the main advantage of the first stage in the proposed framework: benefiting from the computational efficiency of a meta-model and providing a variety of useful results.

Figure 6 presents the results provided by the SPCE-aided MCS with $N_{MC} = 10^5$. According to the obtained 10^5 FoS values, its PDF and CDF can be plotted. The PDF shows that the dam possible FoS under the current calculation configuration mainly varies between 1 and 1.6 with a mean (μ_{FoS}) of 1.285 and a standard deviation (σ_{FoS}) of 0.137. Giving the CDF allows approximately estimating the probability of getting a FoS lower than any threshold. Then, the dam Pf is obtained by computing the ratio between two numbers: N_f and N_{MC} with N_f representing the number of the FoS values lower than 1 (i.e., failure). **Figure 7** presents the results obtained by the SPCE-aided FORM and GSA. The FORM is an approximation method by the fact that it assumes a linear expansion tangent to the LSS at $\boldsymbol{P}^*$ for the Pf estimation.

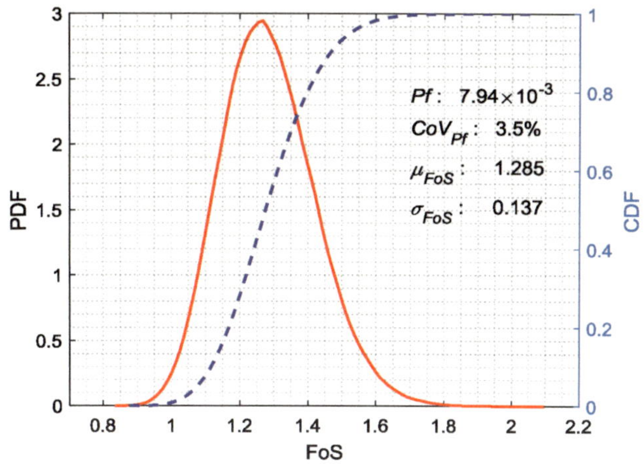

Figure 6.
PDF and CDF of the FoS values obtained in the first stage.

The advantage of the FORM is that it is able to give many results in terms of reliability index (β_{HL}), design point, partial safety factor (FoS) and importance factor of each variable. The design point represents the most probable failure point in the FORM context, and can be used together with the partial FoS to guide a deterministic analysis on the same problem. The GSA aims to quantify the contribution of each soil property, modeled by RV, with respect to the dam FoS variance. The results permit to make a rank of all the variables according to their importance as shown in **Figure 7**. The Kucherenko index is used here for the GSA since there exists a correlation between the input variables. The total effect index considers both the independent impact of one variable and its correlation effect with other variables. According to **Figure 7**, it is observed that the variable ϕ' is dominant for the FoS variation under the current probabilistic input configuration (**Table 2**).

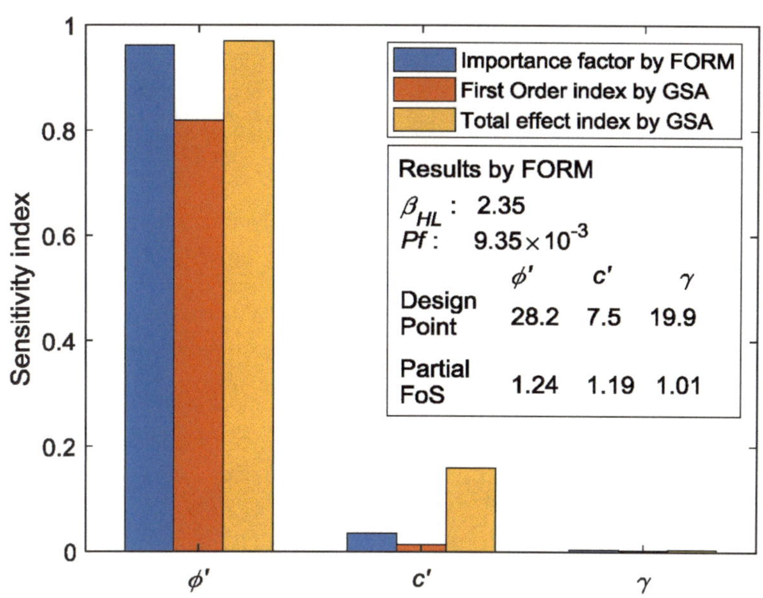

Figure 7.
GSA and FORM results of the dam in the first stage.

The variable c' has also a noticeable contribution while the γ effect is very slight. It should be noted that the importance factor (FORM) and the sensitivity index (Sobol or Kucherenko -based GSA) are different between each other. The former measures the contribution of a RV with respect to the failure while the latter quantifies the importance of a RV regarding the QoI variation. Additionally, the importance factor by FORM holds only for the case with independent RVs. The related results are still given in **Figure 7** in order to have a rank and to compare with the GSA estimates.

In summary, this stage provides a first estimate on the dam *Pf* which can be used to evaluate the design of a new dam or the safety condition of an existing dam. The other information, such as the FoS statistics and design points, are also helpful for this first evaluation. The sensitivity analysis results permit to know the contribution of the considered soil properties and treat their uncertainties with different ways in a next verification/design phase.

4.3 Second stage: RF approach

The second stage of the proposed framework is to consider the soil spatial variability by RFs and obtain a more precise *Pf* estimate. According to the results of the first stage, the effect of the variable γ is almost negligible for the dam failure or the FoS variance. Therefore, it is reasonable to only model c' and ϕ' by RFs and keep representing γ by RVs in the second stage. This can make the analysis of this stage simpler and faster given that generating RFs and mapping them to a model require extra computational efforts. Besides, the input dimension can be reduced compared to considering three RFs (c', ϕ', and γ) for each simulation since there is no need to do the γ discretization. In this stage, the c' and ϕ' are modeled by cross-correlated lognormal RFs using the parameters of **Table 2**, while the γ is treated as same as the previous stage.

The first step in this stage is to determine the truncation term number N_{KL} once the necessary probabilistic parameters (mean, CoV, L_x and L_z) are defined. It can be achieved by evaluating the truncation error of a KLE RF evaluated by Eq. (10) with a target accuracy. In this work, the N_{KL} is determined for a ε_{KL} lower than 5%.

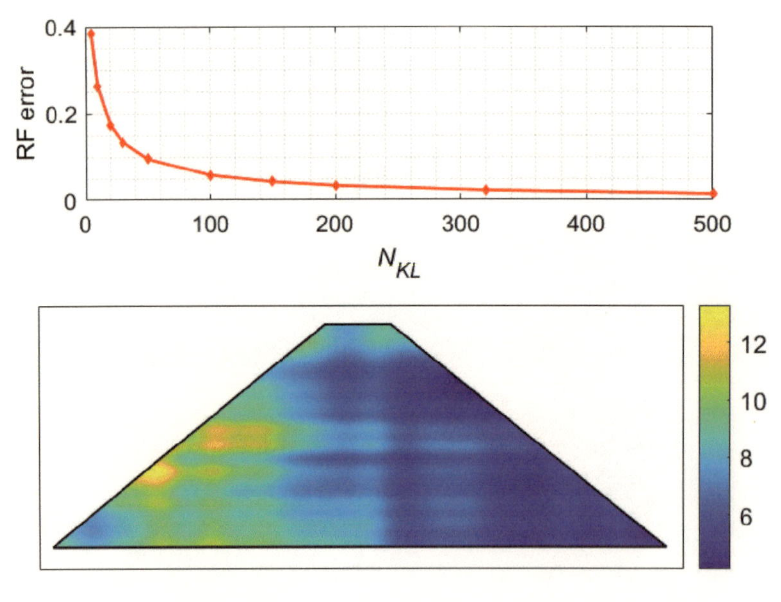

Figure 8.
RF truncation error and example of the generated RFs.

Figure 8 plots the ε_{KL} against the N_{KL} and finally a $N_{KL} = 125$ is adopted for the case of $L_x = 40m$ and $L_z = 8m$ (**Table 2**). Then, the input dimension for the reliability analysis in this stage is 251 since two RFs (c' and ϕ') and one RV (γ) should be considered for each simulation. In **Figure 8**, an example of the c' RF generated by the KLE with the pre-defined parameters is illustrated. It can be seen that c' varies more significantly in the vertical direction that the horizontal one.

The second step is to create an SPCE model to replace the LEM-GA coupled with RFs. The active learning process of **Figure 4** is followed for the SPCE training with the user-defined parameters given as: $Q_t^2 = 0.98$; $Err_t = 0.15$; N_{s2}=5; $N_{add} = 2$; $N_{ini} = 251$; the size for the MCS candidate pool is large enough so that the estimated Pf has a CoV lower than 5%. Additionally, the input dimension is reduced by using the SIR *a priori* the SPCE construction at each iteration with the current DoE. This is because that the considered reliability analysis is a high dimensional problem which has 251 input RVs. Directly training an SPCE with the original input space will require a large size of DoE and may lead to a less accurate meta-model. By performing an SIR with a slice number of 20, the input dimension is reduced from 251 to 19. Then, it is possible to create an SPCE model with respect to the 19 new RVs using an acceptable size of DoE (e.g., several hundred). At the end, the obtained SPCE is a 2-order model with a Q^2 of 0.99. The final size of the DoE is 423 which means that 172 new samples are added in the adaptive process in order to improve the SPCE performance in estimating the dam Pf.

The last step is to perform an MCS with the determined SPCE model. The obtained results are presented in **Figure 9**. The dam FoS mainly varies between 1.1 and 1.5 with a mean of 1.276 and a standard deviation of 0.102. The dam Pf is estimated as 6×10^{-4}. Compared to the analysis of the first stage, the current analysis leads to a clearly reduced σ_{FoS} corresponding to a narrower variation range as shown by the PDF. The dam Pf is also decreased by around one order of magnitude. The comparison between **Figures 6** and **9** indicate that using RFs instead of RVs to model the soil variabilities can reduce the FoS uncertainty and provide a lower Pf estimate. Although considering the soil spatial variability requires extra computational efforts for RFs generation and makes the reliability

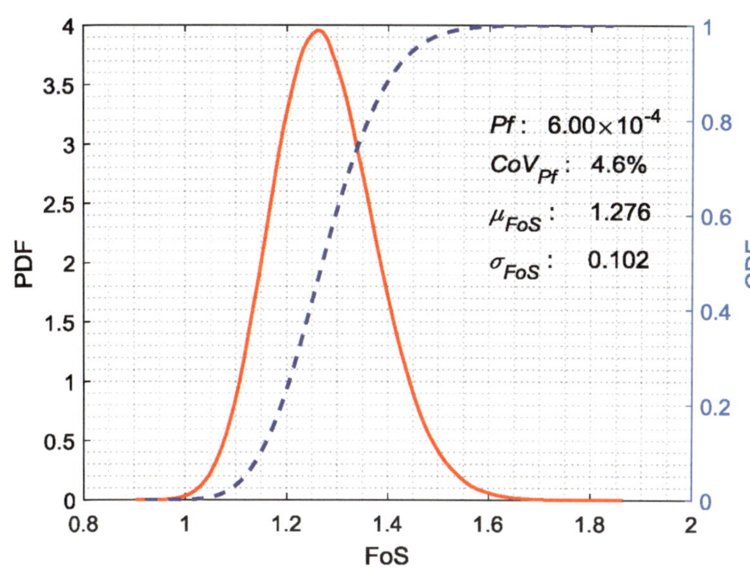

Figure 9.
MCS results of the dam obtained at the second stage.

analysis more complex, it is worthy to do so since a more precise *Pf* estimate can be obtained and can lead to a more economic design. A detailed explanation about the *Pf* decrease from the RV to RF approach will be given later.

4.4 Parametric analysis

It needs in some cases to perform a series of parametric analyses. The objective is to evaluate the effects of some parameters which are difficult to be precisely quantified due to the lack of enough measurements. The physical range recommended in literature for the concerned parameters can be used to define some testing values. In the proposed framework, the computational burden for conducting such parametric analyses is acceptable since the use of the SPCE model significantly reduces the consuming time of one probabilistic analysis. In this work, the effects of two parameters on the dam reliability are investigated: the cross-correlation between c' and ϕ' ($\beta_{c\phi}$) and the vertical autocorrelation distance (L_z). In the reference case (**Table 2**), the $\beta_{c\phi}$ is assumed to be -0.3. In this section, two testing values (0 and $-$ 0.6) are selected for the $\beta_{c\phi}$ to check its influence: $\beta_{c\phi}$ = 0 represents independent input RVs while $\beta_{c\phi}$ = $-$ 0.6 is a strongly correlated case. Then for the L_z, two values (40 and 3 m) are adopted as two complementary cases to the assumed L_z in the reference case (8 m). L_z = 40 m leads to isotropic RFs given that L_x is also 40 m and represents a relatively homogenous soil, while L_z = 3 m allows to consider a soil significantly varying along depth. The L_x is assumed to be constant with 40 m in this case. Such an assumption is based on the fact that embankment dams are usually constructed by layers leading to highly correlated soil properties in the horizontal direction if the construction materials are well selected. **Table 3** gives a summary of all the cases considered in this section. Case 1B and 2B in this table refers to the reference case which is performed respectively in the first and second stage of the previous sections. In the RV approach, the soil is assumed to be homogenous which means that the values of different locations in this field are perfectly correlated. Therefore, this approach corresponds to an infinite L_x and L_z. The input dimension in **Table 3** means the number of input RVs for each case. The dimension is 3 for all the cases with the RV approach which represents the three soil properties (γ, c' and ϕ'). For the RF approach, the dimension is related to the truncation term N_{KL} as determined in **Figure 8**. The N_{KL} should be increased if smaller L_x or L_z are considered. In other words, it means than an accurate representation of a RF with small autocorrelation distances requires more RVs.

Figure 10 presents the obtained results of the parametric analysis (1A, 1B, and 1C) for the $\beta_{c\phi}$ effect. The SPCE is used for the meta-model construction and it is coupled only with MCS since the focus here is to estimate the dam *Pf*. From this figure, it is observed that the FoS PDF becomes taller and narrower when the $\beta_{c\phi}$ is

	Case	Approach	Distribution parameters	$\beta_{c\phi}$	L_x (m)	L_z (m)	Input dimension
$\beta_{c\phi}$ Effect	1A	RV	**Table 2**	0	∞	∞	3
	1B	RV	**Table 2**	−0.3	∞	∞	3
	1C	RV	**Table 2**	−0.6	∞	∞	3
L_z Effect	2A	RF	**Table 2**	−0.3	40	40	61
	2B	RF	**Table 2**	−0.3	40	8	251
	2C	RF	**Table 2**	−0.3	40	3	621

Table 3.
The analyses conducted in this section.

decreased from 0 to -0.6. The PDF of the independent case leads to the most scattered FoS values. Consequently, the dam Pf estimate, being the tail probability of a distribution, is decreased from Case 1A to 1C. The Pf decrease corresponds to a change of one order of magnitude when $\beta_{c\phi}$ is reduced from 0 to -0.6. Considering a negative cross-correlation between c' and ϕ' can reduce the total input uncertainty. Therefore, the output variance can also be reduced given that these two properties are dominant for the FoS variation according **Figure 7**. Additionally, the number of small FoS values is decreased since a negative $\beta_{c\phi}$ can partially avoid generating a small value for both c' and ϕ' in one simulation, which then leads to a lower Pf.

Figure 11 shows the results for the investigation on the L_z effect. The results of Case 1B are presented as well in this figure which permits a comparison between the RV and RF approach. The SPCE-aided MCS is used by following the algorithm of **Figure 4** to perform the reliability analysis. Particularly, the input dimension is reduced by using the SIR each time before the SPCE construction because the three considered cases (2A, 2B, and 2C) are high dimensional problems due to the RF discretization as shown in **Table 3**. It can be observed that the FoS PDF is taller and narrower with decreasing the L_z. This means that a smaller L_z can lead to a FoS with less uncertainty. As a result, the tail probability of the distribution (Pf) is also decreased from Case 2A to 2C with a change of two orders of magnitude showing that the L_z effect is remarkable on the dam Pf. The RV approach provides the most scattered FoS distribution and the highest Pf. A possible explanation for these findings is given as below. Two RFs for respectively Case 2A and 2C are generated and presented in **Figure 12** in order to help the following interpretation. A large L_x or L_z value means a great probability of forming large uniform areas as shown in **Figure 12** (upper part). The global average of the field could be low, medium or high which means a large variation for the global average among different realizations of RFs. The global average is partially related to the estimated FoS so the latter could also have a large variation as evidenced in **Figure 11**. Then, the Pf is higher since it is the tail probability of a distribution. On the contrary, for the case with a

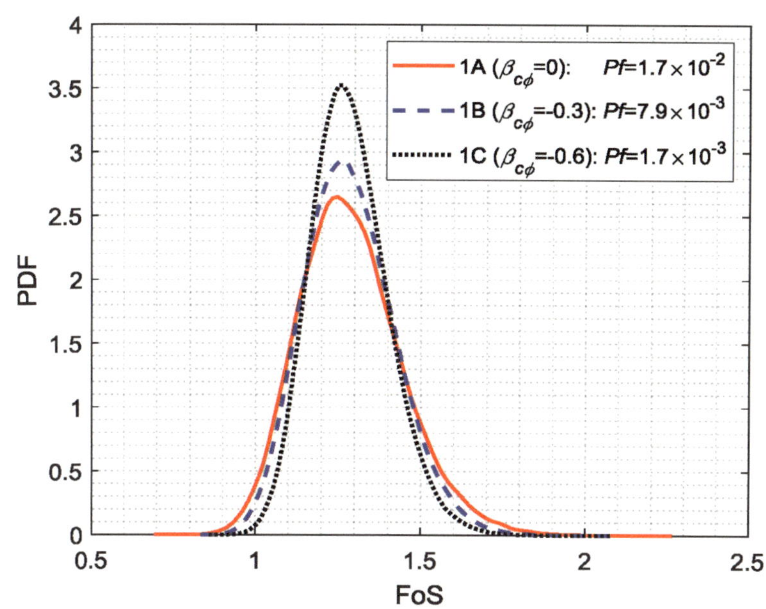

Figure 10.
Comparison of the PDF and Pf by considering different $\beta_{c\phi}$.

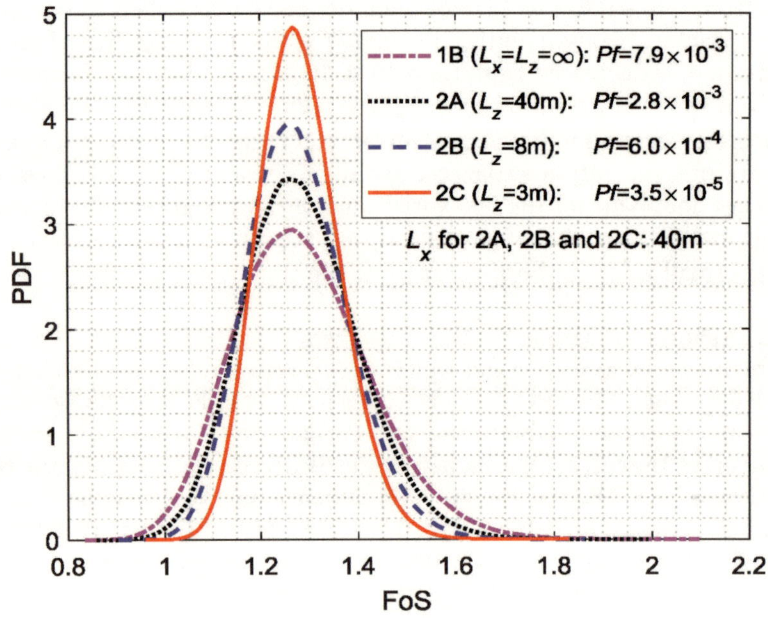

Figure 11.
Comparison of the PDF and Pf considering different L_z.

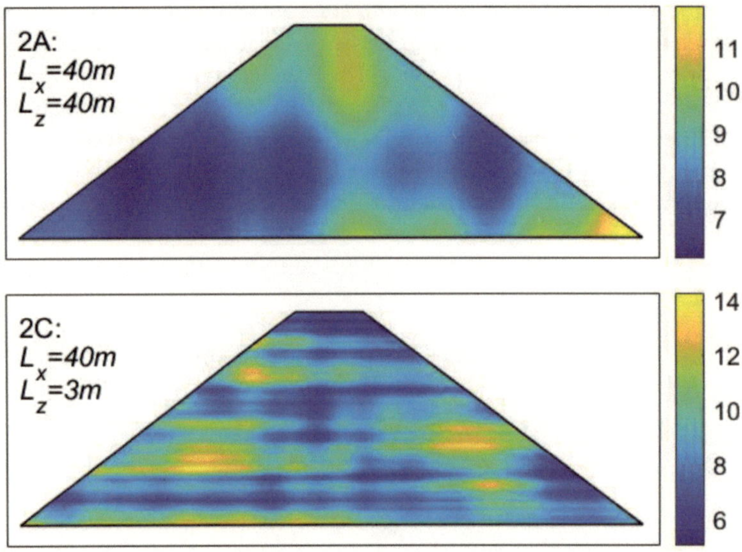

Figure 12.
Example of the RFs generated in Case 2A and 2C.

low L_x or L_z value, there are probably some relatively higher values generated close to the area with low values and vice versa. As a result, the global average varies in a narrower range also the FoS, so the *Pf* is lower. Additionally, the failure surface seeks the weak areas, so it is in general longer and less smooth when L_x and L_z are small. For a long and rough slip surface, more energies are required for its movement which means a relatively high FoS. Therefore, the *Pf* is lower with small L_x and L_z. As these two parameters are assumed to be infinite in the RV approach, the largest uncertainty in the FoS and the highest *Pf* are obtained.

5. Discussions

5.1 Validation of the surrogate-based results

The proposed framework is based on the metamodeling to perform a probabilistic analysis. Therefore, the key element of the proposed framework is to create an accurate SPCE model which can well replace the original computational model. In the next paragraph, two recommendations are given for a good metamodeling.

Firstly, it is recommended to use a space-filling sampling technique (e.g., LHS) to generate samples from a given PDF for the initial DoE and the MCS candidate pool. This allows generating a set of samples which can reasonably cover the input space. The LHS is also faster than a purely random sampling technique for the result convergence in an MCS. Secondly, an active learning process, such as the one of **Figure 4**, is highly suggested for the SPCE construction. The process is stopped only if stable *Pf* estimates are reached and the added samples in this process are those which can improve the SPCE performance in predicting failures. Therefore, one can have more confidence on the obtained *Pf* estimate by using this process. Besides, the DoE is gradually enriched until the stopping conditions are met. As a result, the size of the DoE can be automatically determined, and the issue of overfitting may be avoided.

Concerning the validation of the constructed surrogate model, three solutions are provided here. The first one is to use the available results in the DoE to compute an accuracy indicator for the meta-model, such as the Q^2 in the PCE. The Q^2 is obtained by the leave-one-out error which is a type of the k-fold cross validation techniques. The advantage of this solution is that no complementary model evaluations are required, and the current DoE is fully explored. Then, the second solution is to use a validation set in which new samples, not covered in the current DoE, are generated and evaluated by both the surrogate and deterministic models. The predictions made by the two models for the new samples can be compared in order to check the accuracy of the obtained meta-model. The new samples can be obtained randomly by the LHS or selected close to the LSS so that the meta-model capacity in classifying safe/failure samples is then verified. The third solution involves performing a direct MCS, FORM or GSA to validate the results obtained by the surrogate-aided analyses. Obviously, this solution requires a huge computational effort if a direct MCS or GSA should be conducted which means that no surrogate model is used, and MCS/GSA is directly coupled with the original computational model. Therefore, it is not an applicable solution for all cases. It could be effective when a series of analyses are performed so a direct MCS can be used to validate one analysis.

In this section, the third validation solution is adopted since some parametric analyses are carried out and the employed deterministic model (LEM-GA) is not too time-consuming. Two cases (1B and 2A) are selected for the validation and are analyzed by a direct MCS in this section. The N_{MC} in the direct MCS is determined so that the CoV_{Pf} is around 10%. **Figure 13** compares the FoS PDF of the two analyses obtained by the two methods (SPCE-aided MCS and direct MCS). It clearly shows that the two PDF curves of the two methods are almost superposed with each other for both the two cases. This indicates a good approximation of the SPCE to the original model.

Table 4 gives a detailed comparison between the two methods in terms of *Pf*, FoS statistics and computational efficiency. It is found that the *Pf* of SPCE-MCS is close to the reference result (direct MCS) with an error lower than 6% for both the two cases. The 95% confidence bounds of the *Pf* estimates are also given in this

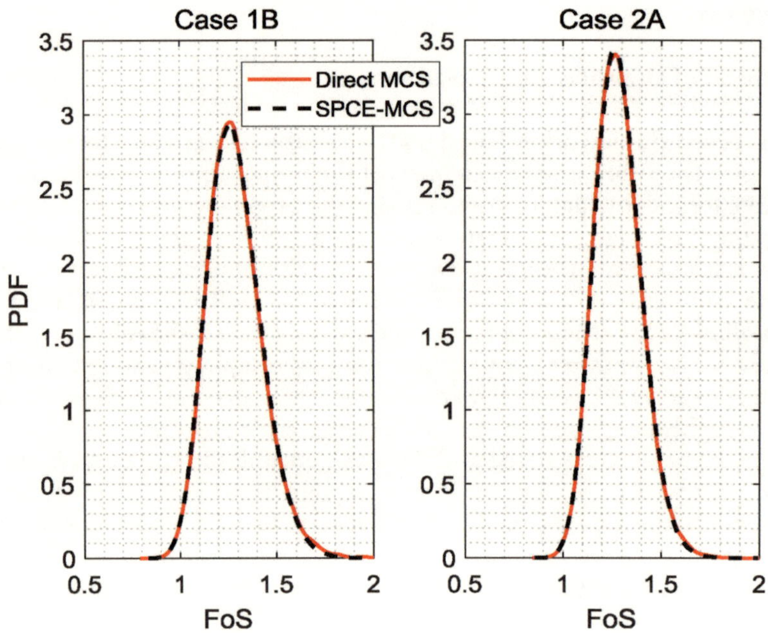

Figure 13.
Comparison of the PDF provided by the two methods.

Case	Method	*Pf* ($\times 10^{-3}$)			FoS		Computational efficiency	
		Mean	Low$^{(1)}$	Up$^{(1)}$	μ_{FoS}	σ_{FoS}	N_{me}	T_{tc}
1B	Direct MCS	7.50	6.07	8.93	1.29	0.14	14,000	25 hours
	SPCE-MCS	7.94	7.39	8.49	1.28	0.14	24	3 minutes
2A	Direct MCS	2.80	2.25	3.35	1.28	0.12	35,000	63 hours
	SPCE-MCS	2.84	2.69	2.99	1.28	0.12	131	25 minutes

$^{(1)}$*95% confidence bounds of the Pf estimate.*

Table 4.
Comparison of the reliability results and computational efficiency provided by the two methods.

table. If the MCS size is large enough, the estimated *Pf* can be approximated by a normal distribution which makes the confidence bounds to be available. It is found that the *Pf* confidence bounds of SPCE-MCS are covered by the ones of the direct MCS. In a surrogate-aided reliability analysis, it is acceptable to largely increase the MCS size in order to obtain a small CoV_{Pf}. However, much more computational efforts are required in the direct MCS if its size should be enlarged, so a CoV_{Pf} of around 10% is adopted in this work. This finding and argument mean that a precise *Pf* with a small CoV_{Pf}could be easily obtained using the proposed framework. Then, for the FoS statistics, the two methods show a good agreement between each other. This is not surprising since the two methods provide closely similar FoS distribution as evidenced in **Figure 13**. Concerning the computational efficiency, two terms are presented in **Table 4** for a comparison: N_{me} (number of deterministic model evaluations) and T_{tc} (total computational time). The T_{tc} is evaluated in a computer equipped with an CPU of Intel Xeon E5-2609 v4 1.7 GHz (2 processors). It is observed that the N_{me} is significantly reduced by using the SPCE compared to a direct MCS (e.g., from 14,000 to 24 for Case 1B), corresponding to a considerable reduction in the T_{tc} (from 25 hours to 3 minutes). Due to the computational efficiency of the SPCE-aided MCS, it is then possible to carry out some parametric analyses in order to investigate the effects of some parameters in a probabilistic framework. The necessary size of the DoE to construct a satisfactory SPCE model is dependent of the input dimension. In general, a higher dimension requires more model evaluations for the SPCE training.

5.2 Practical applications

This section provides a discussion on some issues of a probabilistic analysis. The objective is to help engineers to better implement the proposed framework into practical problems.

5.2.1 Probabilistic analysis tool

Probabilistic analysis has received much attention during the last decade in literature. However, it is still not widely applied in practical engineering problems. One major reason which hinders its application in practice is the complexity of performing a probabilistic analysis including understanding/programming a reliability method, RF generation and couple them with a deterministic model. This problem is being addressed in recent years with the establishment of many probabilistic analysis tools. A variety of reliability/sensitivity methods are available in these tools and can be linked with a computational model developed in a third-party software. Examples of these tools include UQlab in Matlab and OpenTURNS in Python. A review of the structural reliability analysis tools can be found in [15]. Using a well-checked tool to perform the probabilistic analysis of practical engineering problems can also avoid personal programming mistakes which could lead to inaccurate results.

5.2.2 Reliability method selection

The proposed framework is based on the SPCE surrogate model. The SPCE is adopted since it has been widely and successfully used in many studies of geotechnical reliability analysis [9, 13, 16]. Some techniques were proposed to be coupled with SPCE in order to efficiently consider the cases with RFs [17], so the SPCE can also handle high dimensional stochastic problems. However, the proposed framework is not limited to the SPCE. It can be updated by using another metamodeling technique (e.g., Kriging and Support Vector Machine) with some necessary modifications. Besides, for estimating a very low Pf (e.g., $<10^{-6}$), the SPCE-aided MCS could be time-consuming given that generating a great number of samples (e.g., $N_{MC}>10^8$) and operating them requires a big memory in a PC. To tackle this problem, the SPCE can be coupled with other reliability methods in order to alleviate the computational burden. The Subset Simulation (SS) [6, 18] is a good choice to replace the MCS for the above-mentioned case, because SS is independent of the input dimension and LSS complexity.

5.2.3 Parameter selection

This chapter focuses on presenting the proposed framework and showing its application to a dam problem. The soil variability modeling is not explained in detail. How to properly describe the soil uncertainties by using a limited number of measurements is also an important element for geotechnical probabilistic analysis in practice. Some studies on this topic can be found in [10, 11]. In this chapter, the effects of two parameters ($\beta_{c\phi}$ and L_z) on the dam reliability are discussed by performing two parametric analyses. Both of them have a significant influence on the dam *Pf* which is decreased with decreasing $\beta_{c\phi}$ or L_z. It seems then logical to use higher values (e.g., $\beta_{c\phi}$=0 and L_z=40 m) in order to obtain conservative results if their values cannot be precisely quantified. Attention must be paid for the selection of L_z or L_x because some recent studies [19] demonstrate that it may exist a worst L_z or L_x which can lead to the highest *Pf*. Therefore, it is advised to perform a parametric analysis on these parameters in order to avoid unsafe designs.

5.2.4 Extension of the proposed framework

The illustrative example in this chapter is based on the stability problem of a homogeneous embankment dam. The proposed framework can also be easily extended to perform the probabilistic analysis of other problems in dams engineering (rapid drawdown, erosion and settlements) by using an appropriate deterministic model and well determining the input uncertainties. Then, the proposed two stages of RV and RF can be conducted in a hierarchical way. For embankment dams with an earth core or multiple soil layers, the uncertainties should be separately modeled for each zone using different RVs or RFs [17]. It is also important to consider the correlation between the variable properties of different zones by analyzing the available measurements. In case of not enough data, a parametric analysis is recommended in order to have an idea of the unknown correlation structure effect. As embankment dams are artificial rock-filled or earth-filled structures constructed with a careful control, uncertainties at the zone boundaries can be considered as negligible. In natural soils, where stratigraphic boundary uncertainties are expected to exist, the related effects will be noticeable and should be considered.

6. Conclusion

This chapter introduces a framework for the probabilistic analysis of embankment dams. The proposed methodology can also be used for other geotechnical works. The RV and RF approaches are both considered in the framework, corresponding two probabilistic analysis stages. In the first stage, the RV approach is used within three probabilistic techniques (MCS, FORM, and GSA) in order to efficiently provide multiple results which could be beneficial for evaluating a design and guide a further site investigation or a further analysis. The second stage introduces RFs for the purpose of accounting for the soil spatial variability and giving a more precise *Pf* estimate. The metamodeling technique, SPCE, is used in both the two stages aiming to alleviate the total computational burden. Particularly, an active learning process is adopted to construct the required SPCE model. This can further reduce the calculation time of a probabilistic analysis and improve the SPCE accuracy in estimating *Pf*. The proposed framework is applied to an embankment dam stability problem. A variety of interesting results for the dam considering the soil uncertainties are obtained. The results include the *Pf*, FoS statistics/distribution, sensitivity index of each soil property, design point and partial safety factors. The provided results (*Pf* and FoS values) are validated by comparing with a direct MCS. The validation also highlights the efficiency of the introduced reliability method which can reduce the total computational time from several days to less than 1 hour for the two considered cases.

DOI: http://dx.doi.org/10.5772/intechopen.92745

Author details

Xiangfeng Guo[1] and Daniel Dias[1,2,3]*

1 3SR laboratory, Grenoble Alpes University, France

2 Antea Group, Antony, France

3 Hefei University of Technology, China

*Address all correspondence to: daniel.dias@3sr-grenoble.fr

References

[1] ICOLD. World Register of Dams. 2020. Available from: https://www.icold-cigb.org/GB/icold/icold.asp [Accessed: 01 April 2020]

[2] Hariri-Ardebili M-A, Salamon J, Mazza G, Tosun H, Xu B. Advances in dam engineering. Infrastructures. 2020;**5**: 39. DOI: 10.3390/infrastructures5050039

[3] Phoon K-K, Kulhawy FH. Characterization of geotechnical variability. Canadian Geotechnical Journal. 1999;**36**:612-624. DOI: 10.1139/t99-038

[4] Griffiths D, Fenton G. Probabilistic Methods in Geotechnical Engineering. Vienna: Springer Vienna; 2007. DOI: 10.1007/978-3-211-73366-0

[5] Guo X, Dias D. Kriging based reliability and sensitivity analysis – Application to the stability of an earth dam. Computers and Geotechnics. 2020;**120**:103411. DOI: 10.1016/j.compgeo.2019.103411

[6] Bourinet J-M. Reliability Analysis and Optimal Design under Uncertainty Focus on Adaptive Surrogate-Based Approaches. Clermont-Ferrand: Université Clermont Auvergne; 2018

[7] Sudret B. Polynomial chaos expansions and stochastic finite element methods. In: Phoon K-K, Ching J, editors. Risk and Reliability in Geotechnical Engineering. Florida: CRC Press; 2014. pp. 265-300

[8] Sudret B. Global sensitivity analysis using polynomial chaos expansions. Reliability Engineering and System Safety. 2008;**93**:964-979. DOI: 10.1016/j.ress.2007.04.002

[9] Pan Q, Dias D. Sliced inverse regression-based sparse polynomial chaos expansions for reliability analysis in high dimensions. Reliability Engineering and System Safety. 2017; **167**:484–493. DOI: 10.1016/j.ress.2017.06.026

[10] Mouyeaux A, Carvajal C, Bressolette P, Peyras L, Breul P, Bacconnet C. Probabilistic stability analysis of an earth dam by stochastic finite element method based on field data. Computers and Geotechnics. 2018; **101**:34–47. DOI: 10.1016/j.compgeo.2018.04.017

[11] Papaioannou I, Straub D. Learning soil parameters and updating geotechnical reliability estimates under spatial variability–theory and application to shallow foundations. Georisk. 2017;**11**:116–128. DOI: 10.1080/17499518.2016.1250280

[12] Marelli S, Sudret B. An active-learning algorithm that combines sparse polynomial chaos expansions and bootstrap for structural reliability analysis. Structural Safety. 2018;**75**:67–74. DOI: 10.1016/J.STRUSAFE.2018.06.003

[13] Guo X, Dias D, Pan Q. Probabilistic stability analysis of an embankment dam considering soil spatial variability. Computers and Geotechnics. 2019;**113**: 103093. DOI: 10.1016.j.compgeo.2019.103093

[14] Loudière D, Hoonakker M, Le Delliou P. Risque sismique et sécurité des ouvrages hydrauliques. Paris: Ministère de l'Écologie de France; 2014 (in French)

[15] Chehade FEH, Younes R. Structural reliability software and calculation tools: A review. Innovative Infrastructure Solutions. 2020;**5**:1–16. DOI: 10.1007/s41062-020-0282-4

[16] Al-Bittar T, Soubra A-H. Efficient sparse polynomial chaos expansion methodology for the probabilistic analysis of computationally-expensive

deterministic models. International Journal for Numerical and Analytical Methods in Geomechanics. 2014;**38**: 1211–1230. DOI: 10.1002/nag.2251

[17] Guo X, Dias D, Carvajal C, Peyras L, Breul P. A comparative study of different reliability methods for high dimensional stochastic problems related to earth dam stability analyses. Engineering Structures. 2019;**188**:591–602. DOI: 10.1016.j.engstruct.2019.03.056

[18] Li HS, Cao ZJ. Matlab codes of subset simulation for reliability analysis and structural optimization. Structural and Multidisciplinary Optimization. 2016;**54**:391–410. DOI: 10.1007/s00158-016-1414-5

[19] Zhu D, Griffiths DV, Fenton GA. Worst-case spatial correlation length in probabilistic slope stability analysis. Géotechnique. 2018;**69**:85–88. DOI: 10.1680/jgeot.17.t.050

Chapter 7

Probabilistic Slope Stability Analysis for Embankment Dams

Yijiang Zhang, Enyue Ji and Weiwei Xu

Abstract

Slope instability is one of the most common forms of dam failure. The commonly used slope stability analysis methods ignore the uncertainty and randomness of dam materials, which may overestimate the stability of dams. In this chapter, a deterministic slope stability analysis based on strength reduction finite-element method is introduced first. After that, the slope is investigated using simple probabilistic concepts and classical slope stability techniques, and the shear strength is treated as a single random variable. Further, the random finite-element method (RFEM) is shown, in which spatial correlation and local averaging are illustrated in detail. Finally, the RFEM is applied to slope stability risk assessment, and the results can lead to higher probabilities of failure.

Keywords: slope stability, finite element, probabilistic methods, dam failure, risk assessment

1. Introduction

Slope instability is one of the most common forms of dam failure. Traditional slope stability analysis methods mainly depend on deterministic analysis, including limit equilibrium analysis and finite-element (FE) analysis. Equilibrium methods mainly include the ordinary method of slices, Bishop's modified method, force equilibrium methods, Janbu's generalized procedure of slices, Morgenstern and Price's method, and Spencer's method. All the equilibrium methods assume that the soil can be divided into slices, which is an artificial distinction. This assumption is the main characteristic that distinguishes different limit equilibrium methods. The main advantage of equilibrium methods is that they involve relatively simpler calculation, which leads to wide use [1–4].

While the finite element method is another powerful approach for slope stability analysis, it can better reflect the stress–strain relationship of soils than the equilibrium methods. Slope failure in the finite-element model occurs naturally through the area in which the shear strength of the soil is insufficient to resist the shear stresses. There are several advantages of a FE approach to slope stability analysis over traditional limit equilibrium methods: (a) there is no assumption about the shape or location of the failure surface, (b) there are no slices and slice side forces, and (c) the FE method is able to monitor progressive failure up to and including overall shear failure [5, 6].

For a practical slope, not only the stress–strain relationship of soils but also the uncertainty of soil properties should be taken into consideration.

IntechOpen

However, traditional slope stability analysis methods always ignore the uncertainty and randomness of dam materials, which may overestimate the stability of dams. Attention was drawn to probabilistic slope stability analyses [7, 8]. Most probabilistic slope stability analyses continue to use classical slope stability analysis techniques which are mainly based on the equilibrium methods [9–12]. An obvious deficiency of the traditional slope stability methods is that the shape of the failure surface is always fixed; therefore, the failure mechanism is not allowed to look for the most critical path through the soil. Besides, these traditional methods cannot take the importance of spatial correlation and local averaging of statistical geotechnical properties into consideration [13–15].

A more rigorous method, in which nonlinear finite-element methods are combined with random field generation techniques, called the random finite-element method (RFEM), was proposed by Griffiths and Fenton [16]. It can fully account for spatial correlation and averaging and is also a powerful slope stability analysis tool that does not require a priori assumptions relating to the shape or location of the failure mechanism.

In this chapter, a deterministic slope stability analysis based on strength reduction finite-element method is introduced first. After that, the slope is investigated using simple probabilistic methods, including first-order second-moment (FOSM) method, first-order reliability method (FORM), and Monte Carlo method. Further, RFEM is shown, in which spatial correlation and local averaging are illustrated in detail. Finally, the RFEM is applied to slope stability risk assessment, and the results can lead to higher probabilities of failure.

2. Deterministic slope stability analysis

Deterministic slope stability analysis in this chapter is based on FE analysis. The program used is called SLOPE64 [6]. This program is for two-dimensional plane strain analysis. The soil is assumed to follow a linear elastic-perfectly plastic behavior characterized by the Mohr-Coulomb shear failure criterion. In the gravity load generation, the stiffness matrix generation, and the stress redistribution procedure, the program uses eight-node quadrilateral elements with simplified integration (four Gauss points per element). Initially, the soil is assumed to be elastic, and the model generates normal and tangential stresses at all Gauss points in the grid. These stresses are then compared with the Mohr-Coulomb failure criterion. If the stress at a particular Gauss point is within the Mohr-Coulomb failure envelope, it is assumed that the position remains elastic. If the stress is on or outside the failure envelope, it is considered that the point is yielding. The yield stresses are redistributed in the whole grids by the viscoplastic algorithm. Overall shear failure occurs when a sufficient number of Gauss points have yielded to allow a mechanism to develop [5, 6].

2.1 Soil model

The soil model used in this program consists of six parameters, as shown in **Table 1**.

The Mohr-Coulomb failure criterion used in this program can be written as follows:

$$F = \frac{\sigma_1' + \sigma_3'}{2} sin\phi' - \frac{\sigma_1' - \sigma_3'}{2} - c' cos\phi' \qquad (1)$$

ϕ'	Friction angle
c'	Cohesion
ψ	Dilation angle
E'	Young's modulus
ν'	Poisson's ratio
γ	Unit weight

Table 1.
Six parameters for soil model.

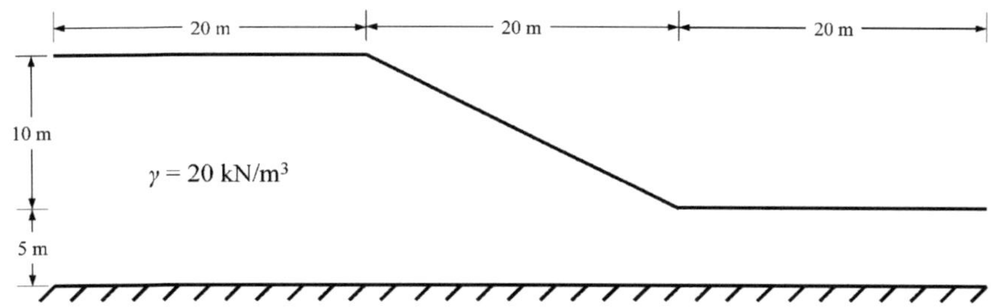

Figure 1.
Homogeneous slope with a foundation layer.

where σ'_1 and σ'_2 are the major and minor principal effective stresses, respectively.

The failure function F can be described as follows:

$F < 0$ stresses inside the failure envelope (elastic).

$F = 0$ stresses on the failure envelope (yielding).

$F > 0$ stresses outside the failure envelope (yielding and must be redistributed).

2.2 Determination of the factor of safety (FS)

The FS of a soil slope is defined as the ratio between the strength of the soils and the actual load. It is exactly the same as that used in traditional limit equilibrium methods. The factored shear strength parameters c_f' and ϕ_f' are therefore given by

$$c'_f = c'/FS$$
$$\phi'_f = \arctan\left(\frac{\tan\phi'}{FS}\right) \tag{2}$$

In this program, in order to find the actual FS, it is necessary to start a systematic search for FS values that will cause the slope to fail. This is achieved by the program that repeatedly solves problems using a sequence of user-specified FS values.

2.3 Slope stability analysis examples

Figure 1 shows a homogeneous slope with a foundation layer. The height of the slope (H) is 10 m, and the thickness of the foundation layer is $H/2$, 5 m. Soil parameters are shown in **Table 2**.

Figure 2 shows the undeformed mesh of the homogeneous slope. The slope is inclined at an angle of 26.578° (2:1). The left boundary is fixed horizontally but is free along the vertical direction, and the base boundary is fixed in both directions. Gravity loads were applied to the mesh, and the trial FS gradually increased until

ϕ'	c'	ψ	E'	ν'	γ
20°	10 kPa	0	10,000 kN/m^2	0.3	20 kN/m^3

Table 2.
Soil parameters.

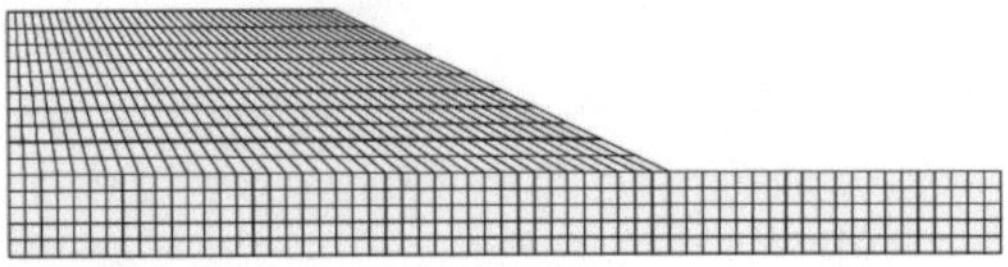

Figure 2.
Undeformed mesh of a homogeneous slope with a foundation layer.

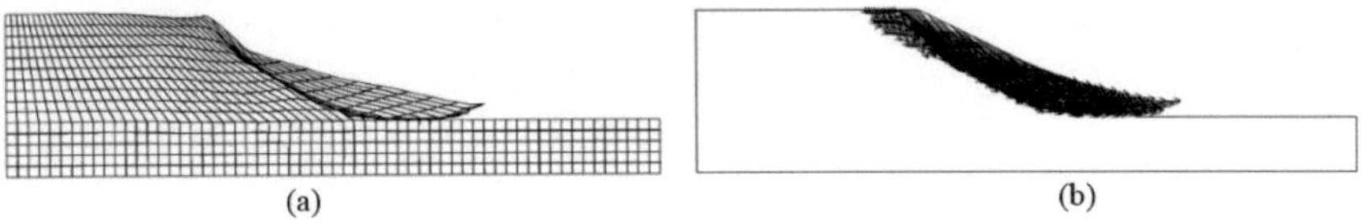

Figure 3.
(a) Deformed mesh of a homogeneous slope with a foundation layer; (b) nodal displacement vectors.

convergence could not be achieved within the iteration limit. The deformed mesh and the nodal displacement vectors are shown in **Figure 3(a)** and **(b)**, respectively. The critical FS is calculated to be 1.34.

3. Classical probabilistic slope stability analysis

In this section, a homogeneous slope and an infinite slope are investigated using simple and classical probabilistic slope stability methods, including FOSM, FORM, and Monte Carlo method. These methods are illustrated one by one in detail followed by a simple example, respectively.

3.1 FOSM

The FOSM method is a relatively simple method of including the effects of variability of input variables on a resulting dependent variable [17, 18]. It is basically a formalized methodology based on a first-order Taylor series expansion. This expansion is truncated after the linear term. The modified expansion is then used, along with the first two moments of the random variable(s), to determine the values of the first two moments of the dependent variable [19–21].

Consider a function f (X, Y) of two random variables X and Y. The Taylor series expansion of the function about the mean values (μ_X, μ_Y) gives

$$f(X,Y) \approx f(\mu_X,\mu_Y) + (X-\mu_X)\frac{\partial f}{\partial x} + (Y-\mu_Y)\frac{\partial f}{\partial y} \tag{3}$$

where derivatives are evaluated at (μ_X, μ_Y).

To a first order of accuracy, the expected value of the function is given by

$$E[f(X,Y)] \approx f(E[X],E[Y]) \tag{4}$$

and the variance by

$$Var[f(X,Y)] \approx Var\left[(X-\mu_X)\frac{\partial f}{\partial x} + (Y-\mu_Y)\frac{\partial f}{\partial y}\right] \tag{5}$$

Hence,

$$Var[f(X,Y)] \approx \left(\frac{\partial f}{\partial x}\right)^2 Var[X] + \left(\frac{\partial f}{\partial y}\right)^2 Var[Y] + 2\frac{\partial f}{\partial x}\frac{\partial f}{\partial y} Cov[X,Y] \tag{6}$$

where E[X] and E[Y] are the expected values of X and Y, respectively; Var[X] and Var[Y] are the variances of X and Y, respectively; Cov[X,Y] is the covariance of X and Y, and Cov[X,Y] = E[(X-E[X])(Y-E[Y])].

If X and Y are uncorrelated,

$$Var[f(X,Y)] \approx \left(\frac{\partial f}{\partial x}\right)^2 Var[X] + \left(\frac{\partial f}{\partial y}\right)^2 Var[Y] \tag{7}$$

In general, for a function of n uncorrelated random variables, the FOSM method tells us that

$$Var[f(X_1, X_2, \dots, X_n)] \approx \sum_{i=1}^{n} \left(\frac{\partial f}{\partial x_i}\right)^2 Var[X_i] \tag{8}$$

where the first derivatives are evaluated at the mean values ($\mu_{X1}, \mu_{X2}, \dots, \mu_{Xn}$).

Here is another example on the homogeneous slope presented in Section 2.3; a probabilistic analysis using FOSM is investigated on this slope. The shear strength parameters are as follows:

$$\mu_{\phi'} = 20°, \sigma_{\phi'} = 3°$$

$$\mu_{c'} = 10kN/m^2, \sigma_{c'} = 3.0kN/m^2$$

According to Eqs. 4 and 7, the expect and variance of FS can be expressed as

$$E[FS] \approx FS(\mu_{\phi'}, \mu_{c'}) \tag{9}$$

$$Var[FS] \approx \left(\frac{\partial(FS)}{\partial \phi\prime}\right)^2 Var[\phi'] + \left(\frac{\partial(FS)}{\partial c\prime}\right)^2 Var[c'] \tag{10}$$

Using a central difference estimate of the derivatives with perturbations of $\pm\sigma$, then

$$Var[FS] \approx \left(\frac{\Delta FS_{\phi\prime}}{2}\right)^2 + \left(\frac{\Delta FS_{c\prime}}{2}\right)^2 \tag{11}$$

where

$$\begin{aligned} \Delta FS_{\phi'} &= FS(\mu_{\phi'} + \sigma_{\phi'}, \mu_{c'}) - FS(\mu_{\phi'} - \sigma_{\phi'}, \mu_{c'}) \\ \Delta FS_{c'} &= FS(\mu_{\phi'}, \mu_{c'} + \sigma_{c'}) - FS(\mu_{\phi'}, \mu_{c'} - \sigma_{c'}) \end{aligned} \tag{12}$$

Using program SLOPE64, FS calculated for each case is shown in **Table 3**.

	ϕ'	c'	FS	
$\mu_{\phi'}, \mu_{c'}$	20.0	10.0	1.34	μ_{FS} = 1.34
$\mu_{\phi'} + \sigma_{\phi'}, \mu_{c'}$	23.0	10.0	1.50	$\Delta FS_{\phi'}$ = 0.3
$\mu_{\phi'} - \sigma_{\phi'}, \mu_{c'}$	17.0	10.0	1.20	
$\mu_{\phi'}, \mu_{c'} + \sigma_{c'}$	20.0	13.0	1.48	$\Delta FS_{c'}$ = 0.28
$\mu_{\phi'}, \mu_{c'} - \sigma_{c'}$	20.0	7.0	1.20	

Table 3.
Factor of safety for five cases.

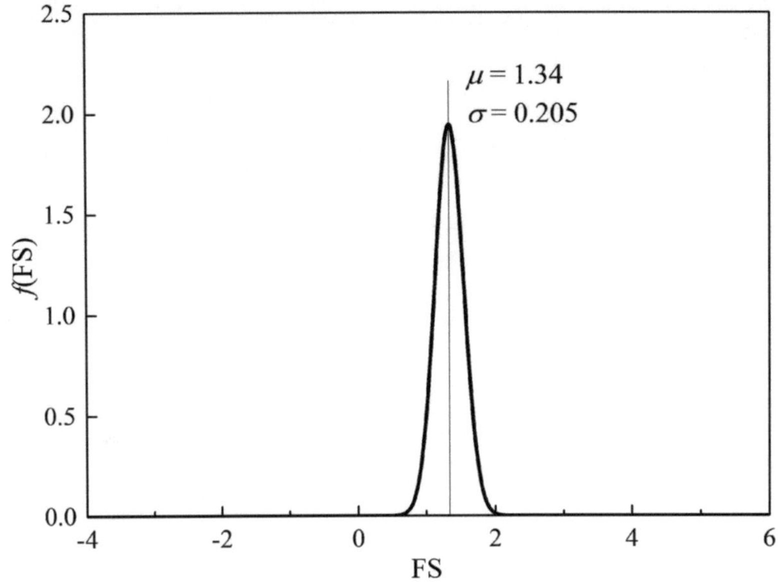

Figure 4.
Normal distribution of FS.

So, the variance of FS can be calculated by

$$Var[FS] = \left(\frac{\Delta FS_{\phi'}}{2}\right)^2 + \left(\frac{\Delta FS_{c'}}{2}\right)^2 = \left(\frac{0.3}{2}\right)^2 + \left(\frac{0.28}{2}\right)^2 = 0.0421$$

Hence

$$\sigma_{FS} = \sqrt{0.0421} = 0.205$$

Assume that the FS probability density function is normal distribution (as shown in **Figure 4**).

$$p[FS < 1] = \Phi\left(\frac{1 - 1.34}{0.205}\right) = \Phi(-1.66) = 1 - \Phi(1.66) = 1\text{-}0.9515 = 0.0485(4.85\%)$$

Consider a "performance function" for this problem in which failure is defined when $M < 0$, the reliability index β in this case is given by $\beta = \frac{E[M]}{\sqrt{Var[M]}}$.

There are three different approaches calculating the reliability index β listed as follows.

3.1.1 Approach 1

For nonnegative loads and resistances (typical in geotechnical engineering), an alternative definition of the performance function could be

$$M = \frac{R}{Q} - 1 \tag{13}$$

so that failure occurs when $M < 0$ as before.

Once more assuming R and Q are uncorrelated, the FOSM method gives

$$E(M) = \frac{E[R]}{E[Q]} - 1 \tag{14}$$

$$\begin{aligned} Var[M] &\approx \left(\frac{\partial M}{\partial R}\right)^2 Var[R] + \left(\frac{\partial M}{\partial Q}\right)^2 Var[Q] \\ &= \frac{1}{E^2[Q]} Var[R] + \frac{E^2[R]}{E^4[Q]} Var[Q] \end{aligned} \tag{15}$$

Hence $\beta = \frac{\mu_Q(\mu_R - \mu_Q)}{\sqrt{\mu_Q^2\sigma_R^2 + \mu_R^2\sigma_Q^2}}$.

3.1.2 Approach 2

In the classical "resistance" versus "load" problem, the performance function can be defined as

$$M = R - Q \tag{16}$$

Assuming R and Q are uncorrelated, the FOSM method gives

$$E[M] = E[R] - E[Q] = \mu_R - \mu_Q \tag{17}$$

And

$$Var[M] = \left(\frac{\partial M}{\partial R}\right)^2 Var[R] + \left(\frac{\partial M}{\partial Q}\right)^2 Var[Q] = \sigma_R^2 + \sigma_Q^2 \tag{18}$$

Hence $\beta = \frac{\mu_R - \mu_Q}{\sqrt{\sigma_R^2 + \sigma_Q^2}}$ which is obviously different to Approach 1.

3.1.3 Approach 3

For nonnegative loads and resistances (typical in geotechnical engineering), an alternative definition of the performance function could be

$$M = \ln\left(\frac{R}{Q}\right) \tag{19}$$

so that failure occurs when $M < 0$ as before.

Once more assuming R and Q are uncorrelated, the FOSM method gives

$$E[M] = \ln\left(\frac{E[R]}{E[Q]}\right) = \ln\mu_R - \ln\mu_Q \tag{20}$$

$$\begin{aligned} Var[M] &\approx \left(\frac{\partial M}{\partial R}\right)^2 Var[R] + \left(\frac{\partial M}{\partial Q}\right)^2 Var[Q] \\ &= \frac{Var[R]}{R^2} + \frac{Var[Q]}{Q^2} \\ &= \frac{Var[R]}{E^2[Q]} + \frac{Var[Q]}{E^2[Q]} \\ &= \nu_R^2 + \nu_Q^2 \end{aligned} \tag{21}$$

Hence $\beta = \frac{ln\,(\mu_R) - ln\,(\mu_Q)}{\sqrt{\nu_R^2 + \nu_Q^2}}$ which is clearly different to the results before.

Apparently, the reliability index β differs with the definition of the performance function, which is one of the major drawbacks of FOSM. Also, the method takes no account of the form of the probability density function describing the random variables, using only their mean and standard deviation, which ignores the effect of distribution of random variables to the results.

3.2 FORM

The major drawback to the FOSM method when used to compute probabilities relating to failure is that it can give different failure probabilities for the same problem [19, 22], which caused Hasofer and Lind to develop an improved approach, FORM [23].

As shown before, the reliability index β is given as

$$\beta = \frac{E[M]}{\sqrt{Var[M]}} \tag{22}$$

which measures how far the mean of the safety margin M is from zero (assumed to be the failure point) in units of number of standard deviations. The interesting point is on the probability that failure occurs, that is, $M < 0$. Therefore, a unique relationship between the reliability index (β) and the probability of failure (p_f) is given by

$$p_f = 1 - \Phi(\beta) \tag{23}$$

where Φ is the standard normal cumulative distribution function. The point, line, or surface defined by $M = 0$ is called the failure surface.

The inconsistency of the FOSM method is due to that different definitions of margin M may have different mean estimates and different first derivatives. What the FOSM method does is calculating the distance from the average point to the failure surface in the gradient direction of the average point [18]. Hasofer and Lind solved the inconsistent problem by looking for the overall minimum distance between the average point and the failure surface, rather than just along the gradient direction [23].

In the general case, suppose that the safety margin M is a function of a sequence of random variables $X^T = \{X_1, X_2, \ldots\}$, that is, $M = f(X_1, X_2, \ldots)$, and that the random variables $X_1, X_2, \ldots$ have covariance matrix C. Then, the Hasofer-Lind reliability index β is defined by [23].

$$\beta = \min_{M=0} \sqrt{(x - E[X])^T C^{-1} (x - E[X])} \tag{24}$$

which is the minimum distance between the failure surface ($M = 0$) and the mean point (E [X]) in units of number of standard deviations. For example, if $M = f$ (X), then Eq. (24) simplifies to $\beta = min_x (x - \mu_X)/\sigma_X$. It is an iterative process to find β under this definition. On the curve $M = 0$, choose a value of x_0 and compute β_0, choose another point x_1 on $M = 0$ and compute β_1, and so on. The Hasofer-Lind reliability index is the minimum of all possible values of β_i. When the minimum reliability index β is determined, the probability of failure can be calculated by Eq. (23).

Figure 5 gives an example for an infinite slope. In this example, $H = 5$ m, $\gamma = 20$ kN/m^3, $\alpha = 30°$, c' and $\tan\phi'$ are lognormal random variables with $\mu_c' = 10$ kPa, $\sigma_c' = 3$ kPa ($\nu_c' = 0.3$) and $\phi' = 30°$, $\mu_{\tan\phi'} = 0.5774$, $\sigma_{\tan\phi'} = 0.1732$ ($\nu_{\tan\phi'} = 0.3$); the logarithmic normal distributions of c' and $\tan\phi'$ are shown in **Figure 6**.

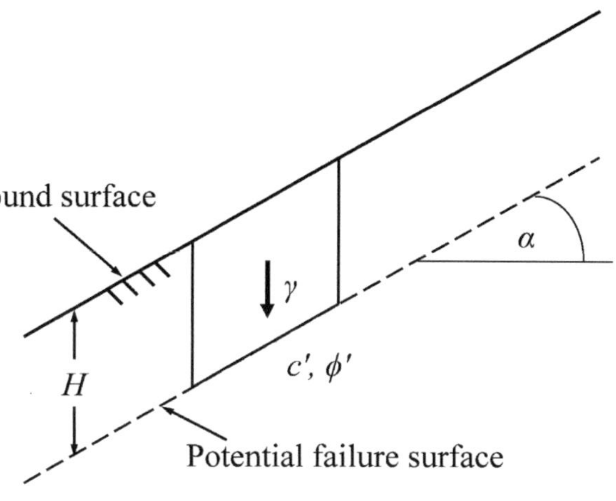

Figure 5.
Infinite slope.

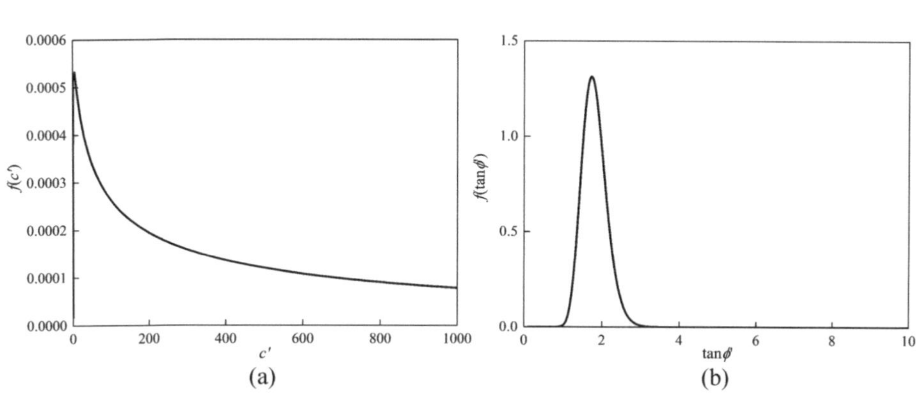

Figure 6.
Logarithmic normal distributions of (a) c′ and (b) tanϕ′.

FS for this slope can be expressed as follows [24]:

$$FS = \frac{c'}{\gamma H \sin\alpha \cos\alpha} + \frac{\tan\phi'}{\tan\alpha} \tag{25}$$

where H is the height of the slope, γ is the saturated unit weight, α is the slope angle to the horizontal direction, c' is the effective cohesion, and ϕ' is the effective friction angle.

Using Eq. (25), it can be calculated that $\overline{FS} = 1.23$. Further, according to FORM algorithm, $\beta = 0.835, p_f = 20.20\%$.

In practical applications, there are many different complex optimization algorithms, usually involving the gradient of M, which can find the point where the failure plane is perpendicular to the origin. The distance between these two points is β [25, 26]. Now, many spreadsheet programs include algorithms that allow users to specify only the minimum equations and constraints on the solution. Unfortunately, nonlinear failure surfaces can sometimes have multiple local minima, with respect to the mean point, which further complicates the problem. In this case, techniques such as simulated annealing may be necessary, but which still do not guarantee finding the global minimum. Monte Carlo simulation is an alternative means of computing failure probabilities which is simple in concept. Furthermore, it is not limited to first order and can be extended easily to very difficult failure problems with only a penalty in computing time to achieve a high level of accuracy [16].

3.3 Monte Carlo method

The Monte Carlo method is a broad computational algorithm that relies on repeated random sampling to obtain numerical results. The basic concept is to use random numbers (sometimes pseudo-random numbers) to solve problems that might be deterministic in principle. This method was proposed in the 1940s and has been widely used in slope stability probability analysis [12, 27–29].

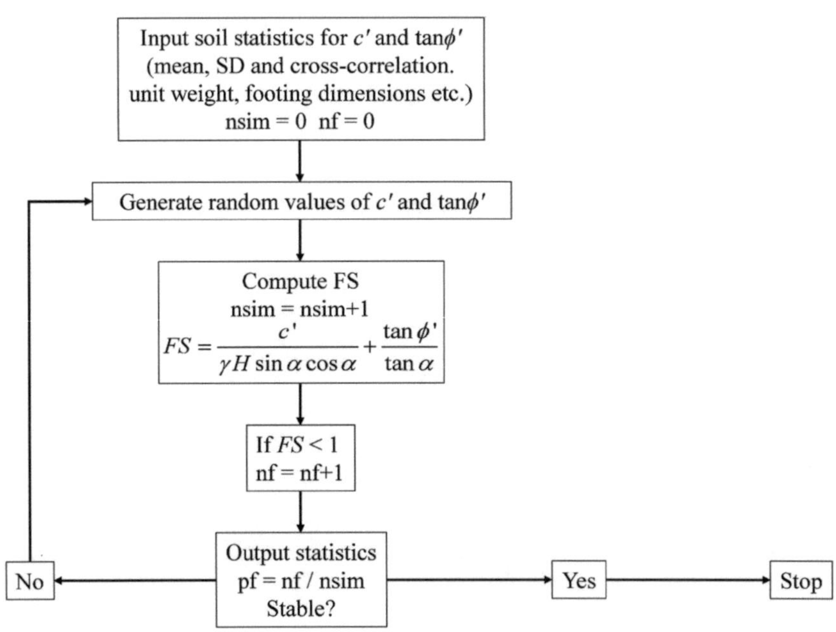

Figure 7.
Algorithm for Monte Carlo analysis of slope stability.

The idea of the Monte Carlo method is to randomly generate samples according to an input probability density function and evaluate the model response of each sample by a deterministic computational model. Consider the problem of determining the probability of failure of a system which has two random inputs, X_1 and X_2. The response of the system to these inputs is then defined as a function $g(X_1, X_2)$. Obviously, the function $g(X_1, X_2)$ is also random because the input variables are random. Assume that system failure will occur when $g(X_1, X_2) > g_{crit}$, where g_{crit} represents the critical state. In the space of (X_1, X_2) values, there will be some region in which $g(X_1, X_2) > g_{crit}$, and the problem boils down to assessing the probability that the particular (X_1, X_2) which actually occurs will fall into the failure region. So the probability p_f can be defined as

$$p_f = P[g(X_1, X_2) > g_{crit}] \tag{26}$$

Figure 7 shows the algorithm for Monte Carlo analysis of slope stability.

Consider the same infinite slope given in **Figure 5**, c' and $\tan\phi'$ are lognormal random variables with $\mu_c' = 10$ kPa, $\sigma_c' = 3$ kPa ($\nu_c' = 0.3$) and $\phi' = 30°$, $\mu_{\tan\phi'} = 0.5774$, $\sigma_{\tan\phi'} = 0.1732$ ($\nu_{\tan\phi'} = 0.3$), which are the same with the previous example. It can be calculated that $\overline{FS} = 1.23, p_f = 23.6\%$. Compared with the probability of failure calculated by FORM, p_f calculated using the Monte Carlo method is a little higher.

4. RFEM slope stability analysis

In this part, a new parameter spatial correlation and the local averaging method are illustrated first. After that, random finite-element method is presented. Finally, results from a full RFEM method are analyzed. Throughout this section, the probability of failure (p_f) is compared with the traditional FS that would be obtained from charts or classical limit equilibrium methods.

4.1 Spatial correlation

In probabilistic slope stability study, the shear strength c and ϕ are assumed to be characterized statistically by a normal distribution or lognormal distribution defined by means μ_c and $\mu_{\tan\phi}$ and standard deviations σ_c and $\sigma_{\tan\phi}$. The probability of the strength that is less than a given value can be found from standard normal distribution table. When the variables are characterized by lognormal distribution, the lognormal can be transformed to normal as follows (take c for example):

$$P[c < a] = P[\ln c < \ln a] = P\left[Z < \frac{\ln a - \mu_{\ln c}}{\sigma_{\ln c}}\right] = \Phi\left(\frac{\ln a - \mu_{\ln c}}{\sigma_{\ln c}}\right) \tag{27}$$

The lognormal parameters $\mu_{\ln c}$ and $\sigma_{\ln c}$ given μ_c and σ_c are obtained via the transformations:

$$\begin{aligned} \sigma_{\ln c}^2 &= \ln\left(1 + \nu_c^2\right) \\ \mu_{\ln c} &= \ln\left(\mu_c\right) - \frac{1}{2}\sigma_{\ln c}^2 \end{aligned} \tag{28}$$

in which the coefficient of variation of c, ν_c, is defined as

$$\nu_c = \frac{\sigma_c}{\mu_c} \tag{29}$$

Unlike the former simulation, another parameter, the spatial correlation length θ_c or $\theta_{\ln c}$, will be considered in the following study. The spatial correlation length describes the significant correlation distance between spatially random values in the Gaussian field. Thus, a small value of θ refers to a ragged field, while a large value refers to a smooth field. In practice, the spatial correlation length can be estimated from a set of shear strength data (c and ϕ) taken over some spatial region simply by performing the statistical analyses on the data.

It has been suggested that typical ν_c values for undrained shear strength lie in the range of 0.1–0.5. The spatial correlation length, however, is less well documented and may well exhibit anisotropy, especially when soils are typically horizontally layered. To simplify in this chapter, the spatial correlation will be assumed to be isotropic.

4.2 Local averaging

The local average subdivision (LAS) method is a fast and accurate method that produces realizations of a discrete local average random process [30]. Consider a random process Z; **Table 4** presents the local average procedure via the LAS method. The algorithm proceeds as follows:

1. Generate a normally distributed random number Z_1^0 with mean zero; the variance is obtained from the random field.

2. Subdivide Z_1^0 into two equal parts, Z_1^1 and Z_2^1; the means and variances should be satisfied with three criteria:

    a. Their variances meet the requirements of local averaging theory.

    b. The relationship between Z_1^1 and Z_2^1 meets the requirements of local averaging theory.

    c. The means of Z_1^1 and Z_2^1 are equal to the mean of Z_1^0, that is, $Z_1^0 = \frac{1}{2}\left(Z_1^1 + Z_2^1\right)$.

3. Subdivide each cell in stage 1 into another two equal parts; the means and variances should be satisfied with the above three criteria, and another new requirement, Z_1^2 and Z_2^2, should be properly correlated with Z_3^2 and Z_4^2.

4. The above steps are repeated, and the cell is subdivided gradually until the size of the subunit reaches the expected requirement.

Stage 0				Z_1^0				
Stage 1			Z_1^1				Z_2^1	
Stage 2		Z_1^2		Z_2^2		Z_3^2		Z_4^2
Stage 3	Z_1^3	Z_2^3	Z_3^3	Z_4^3	Z_5^3	Z_6^3	Z_7^3	Z_8^3
Stage 4								

Table 4.
Procedure of the LAS method.

Using the RFEM approach to analyze a slope, each element is assigned a constant property, including the mean, standard deviation, and spatial correlation length of the shear strength, at each realization of the Monte Carlo process. The assigned property represents an average over the area of each finite element used to discretize the slope. If the point distribution is normal, local arithmetic averaging is used which results in a reduced variance but the mean is unaffected. In a lognormal distribution, however, local geometric averaging is used, and both the mean and the standard deviation are reduced by this form of averaging as is appropriate for situations in which low-strength regions dominate the effective strength. The reduction in both the mean and standard deviation is from

$$\mu_X = E[X] = e^{\mu_{\ln X} + \frac{1}{2}\sigma^2_{\ln X}} \tag{30}$$

$$\sigma^2_X = Var[X] = \mu^2_X \left(e^{\sigma^2_{\ln X}} - 1\right) \tag{31}$$

The mean of a lognormally random variable depends on both the mean and the variance of the underlying normal log variable:

$$\sigma^2_{\ln X} = \ln\left(1 + \frac{\sigma^2_X}{\mu^2_X}\right) \tag{32}$$

$$\mu_{\ln X} = \ln(\mu_X) - \frac{1}{2}\sigma^2_{\ln X} \tag{33}$$

Obviously, local averaging has a great influence on the form of a reduced mean and standard deviation. These adjustments are fully accounted for in the following RFEM analysis.

4.3 Random finite-element method

A powerful and general method of accounting for spatially random shear strength parameters and spatial correlation is the RFEM, which combines elastoplastic finite-element analysis with random field theory generated using the LAS method. **Figure 8** shows a typical finite-element mesh for the test problem considered in this section. Most of the elements are square, and the elements adjacent to the slope are degenerated into triangles. Taking full account of element size in the local averaging process, the random field of shear strength values was generated and mapped onto the finite-element mesh. In a random field, the value assigned to each finite element is a random variable. The random variables can be correlated to one another by controlling the spatial correlation length $\theta_{\ln c}$ as described previously. **Figure 9a, b**, and **c** shows the typical meshes corresponding to different spatial correlation lengths. **Figure 9a** shows a relatively low spatial correlation length of $\theta = 1$, **Figure 9b** shows a medium spatial correlation length of $\theta = 5$, and **Figure 9c** shows a relatively high spatial correlation length of $\theta = 10$. In these figures, light regions represent weak- or low-strength soils, while dark regions

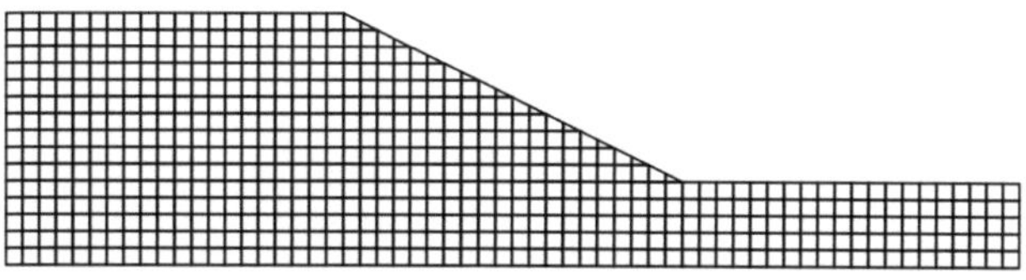

Figure 8.
Undeformed mesh of a homogeneous slope with a foundation layer.

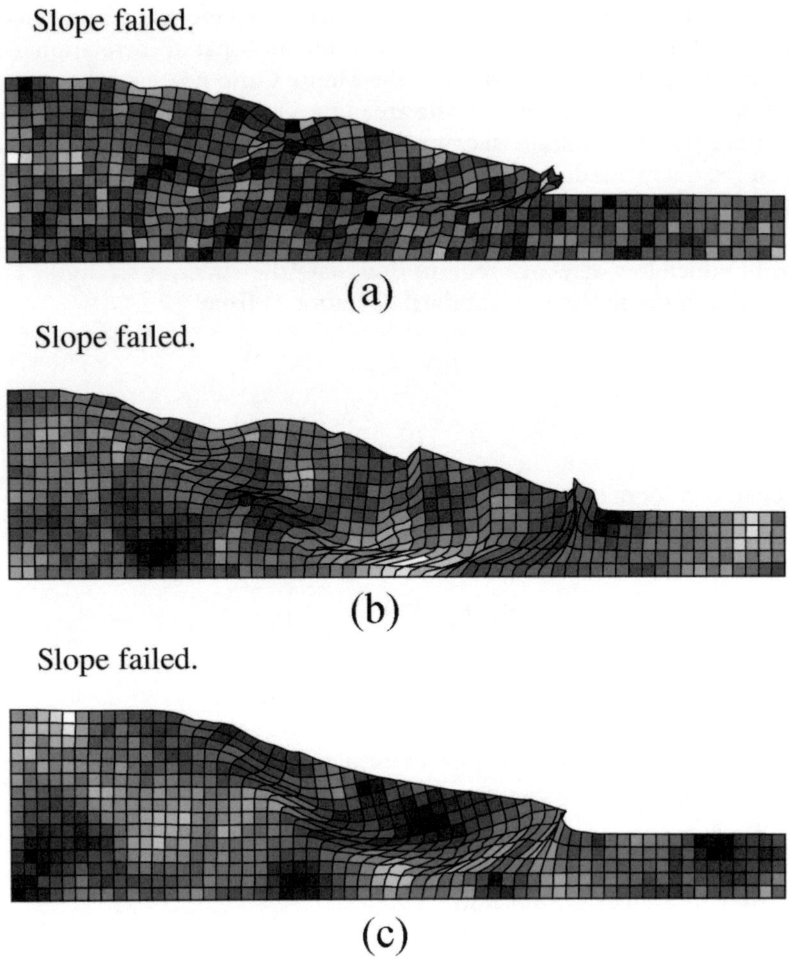

Figure 9.
Deformed mesh at slope failure for three different spatial correlation lengths. (a) θ = 1; (b) θ = 5; (c) θ = 10.

represent strong- or high-strength soils. The shear strength distributions of these three cases come from the same lognormal distribution, and the only difference is the spatial correlation length. The slope stability analyses use the Tresca failure criterion which is an elastic-perfectly plastic stress–strain law. When the stresses exceed the yield stress, the program attempts to redistribute excess stresses to neighboring elements that still have reserves of strength. This process is iterative and will continue until the Tresca criterion and global equilibrium are satisfied at all points within the mesh under quite strict tolerances. Plastic stress redistribution is accomplished using a viscoplastic algorithm with eight-node quadrilateral elements and reduced integration in both the stiffness and stress redistribution parts of the algorithm [5, 6].

4.4 Results of RFEM analyses

Figure 9 shows three typical random field realizations and the associated failure mechanisms for slopes with $\theta = 1$, 5, and 10. It can be concluded that spatial correlation length has a great influence on the failure surface morphology. When θ is low, the shear strength between neighbored elements varies severely; when θ is high, similar properties can be found between neighbored elements. In the RFEM approach, the failure mechanism is free to seek out the weakest path through the soil. Thus, the failure surface will tend to pass through elements with weaker shear strength.

In the following part, the two parameters of shear strength, c and ϕ, are defined as the random variable, respectively, to investigate the influence of spatial correlation length and coefficient of variance on the probability of failure.

4.4.1 Define c *as random*

Defining friction angle as a deterministic parameter, ϕ = 20°, and then fixing the mean of cohesion with μ_c = 10 kPa, **Figure 10** shows the probability of failure p_f as a function of the spatial correlation length θ for a range of coefficients of variation, with the mean cohesion fixed at μ_c = 10 kPa. **Figure 11** shows the relationship between probability of failure p_f and the coefficient of variation ν_c with two different spatial correlation lengths. It can be seen from **Figure 10** that the probability of failure can be divided into two branches, with the probability of failure tending to unity or zero for higher and lower values of ν_c, respectively. **Figure 11** demonstrates that when θ becomes large, the probability of failure is overestimated (conservative) when the coefficient of variation is relatively small and underestimated (unconservative) when the coefficient of variation is relatively high. The RFEM results show that the inclusion of spatial correlation and local averaging in this case will always lead to a smaller probability of failure.

4.4.2 Define ϕ *as random*

Defining cohesion as a deterministic parameter, c = 10 kPa, and then fixing the mean of friction angle with μ_ϕ = 20°, **Figures 12** and **13** show the effect of the spatial correlation length θ and the coefficient of variation ν_ϕ on the probability of failure for the test problem. It is obvious that **Figures 12** and **13** show similar tendency with **Figures 10** and **11**. Comparing **Figures 11** and **13**, it can be concluded that the influence of spatial correlation length of ϕ on the probability of failure is less than that of c.

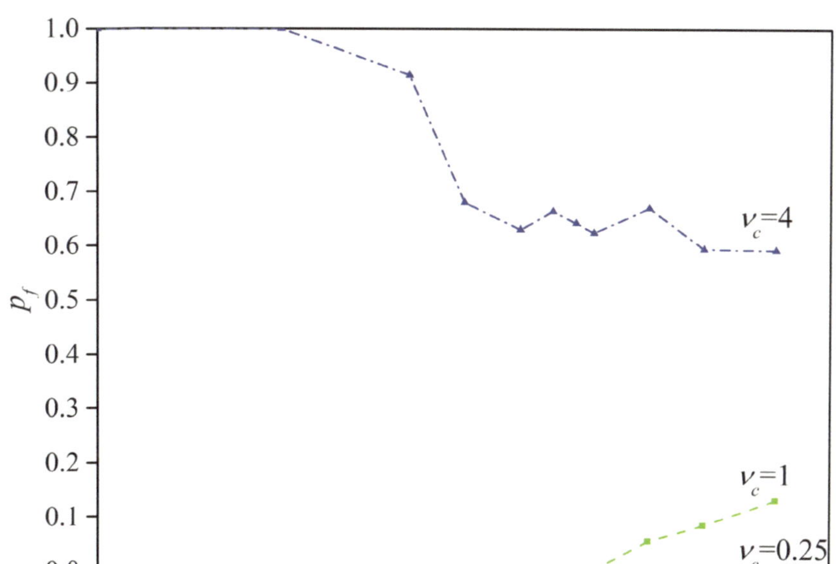

Figure 10.
Probability of failure versus spatial correlation length (the mean of cohesion is fixing at μ_c = 10 *kPa).*

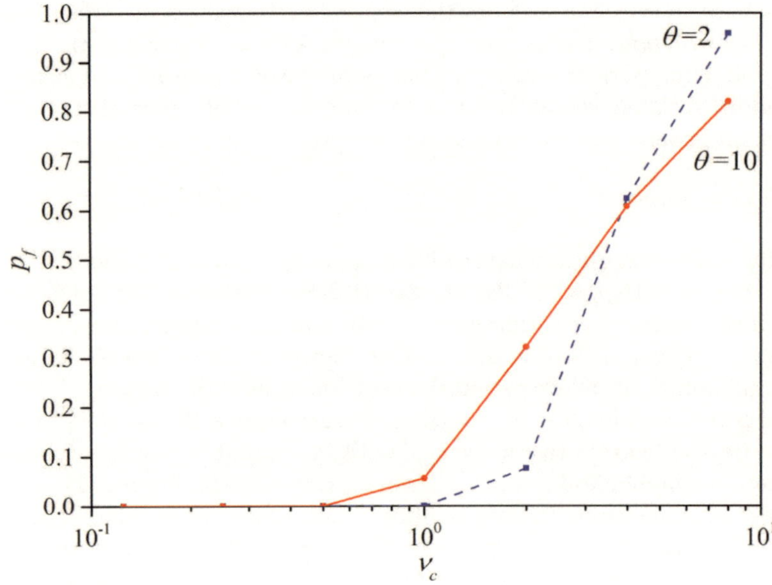

Figure 11.
Probability of failure versus coefficient of variance (the mean of cohesion is fixing at μ_c = *10 kPa).*

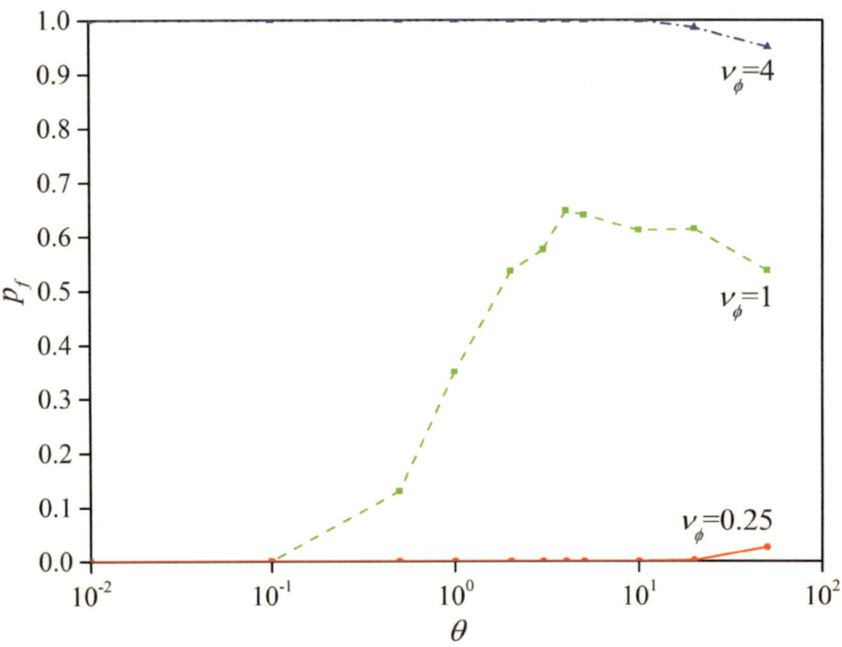

Figure 12.
Probability of failure versus spatial correlation length (the mean of friction angle is fixing at μ_ϕ = *20°).*

4.4.3 Define c *and* ϕ *as random*

Defining cohesion c and friction angle ϕ as random parameters, and then fixing the mean of cohesion with μ_c = 10 kPa and the mean of friction angle with μ_ϕ = 20°, **Figure 14** shows the probability of failure versus spatial correlation length with different coefficients of variance of c and ϕ. Clearly, **Figure 14** shows similar tendency with **Figures 10** and **12**. **Figure 15** shows the probability of failure p_f as a

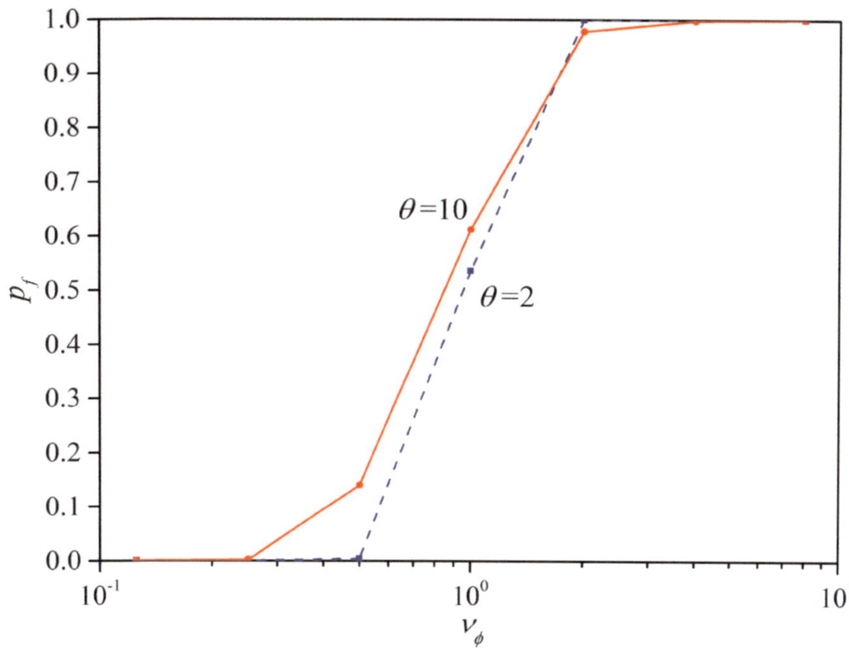

Figure 13.
Probability of failure versus coefficient of variance (the mean of friction angle is fixing at μ_ϕ = 20°*).*

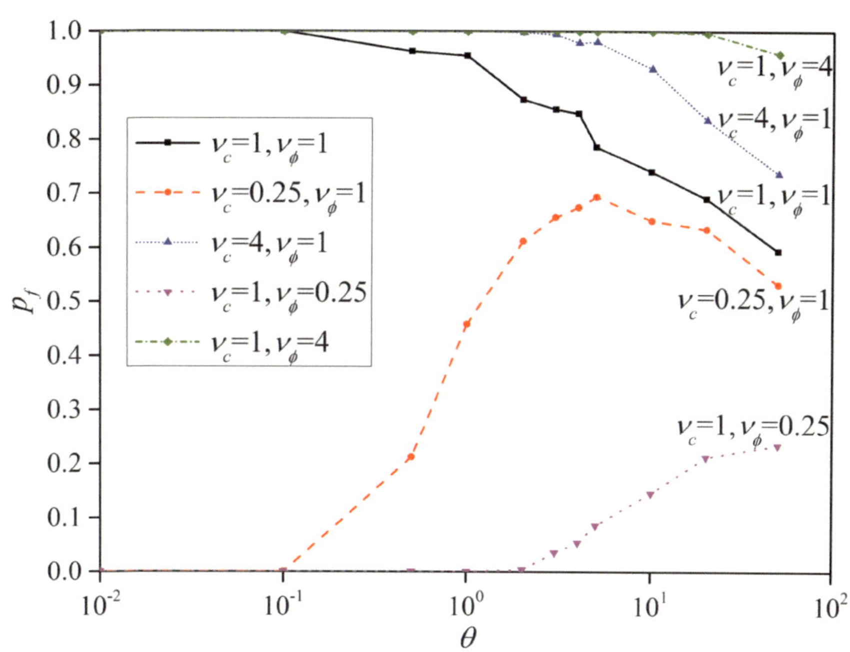

Figure 14.
Probability of failure versus spatial correlation length (the mean of cohesion is fixing at μ_c = 10 *kPa and the mean of friction angle is fixing at* μ_ϕ = 20°*).*

function of coefficient of variance ν_c for two different θ = 2 and 10, with the mean cohesion fixed at μ_c = 10 kPa, the mean of friction angle fixing at μ_ϕ = 20° and ν_ϕ fixing at 1. Similarly, **Figure 16** shows the probability of failure p_f as a function of coefficient of variance ν_ϕ for two different θ = 2 and 10, with the mean cohesion

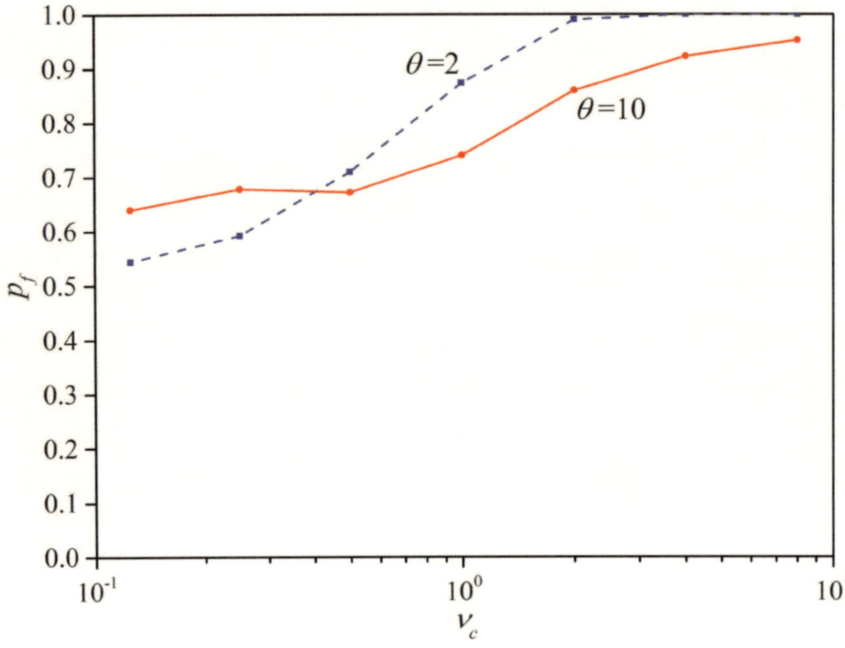

Figure 15.
Probability of failure versus coefficient of variance of cohesion (the coefficient of variance of friction angle is fixing at ν_ϕ = 1).

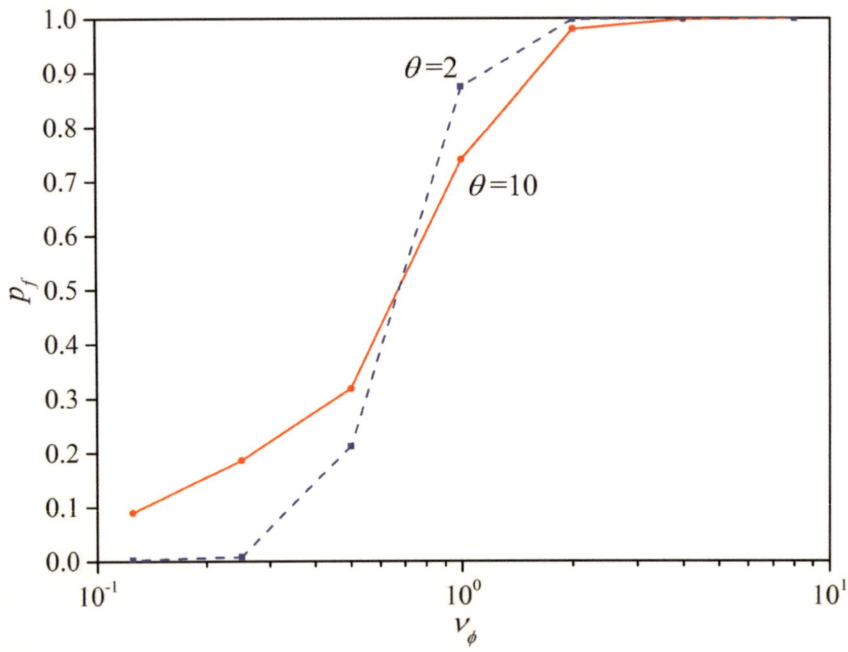

Figure 16.
Probability of failure versus coefficient of variance of friction angle (the coefficient of variance of friction angle is fixing at ν_c = 1).

fixed at μ_c = 10 kPa, the mean of friction angle fixing at μ_ϕ = 20° and ν_c fixing at 1. Clearly, these two figures show a similar relationship with **Figures 11** and **13**. It is worth noting that defining ϕ as random has an apparent influence on the probability of failure versus coefficient of variance of cohesion. From **Figure 15**, p_f is relatively higher than the case that only c is the only random parameter.

5. Conclusion

This chapter presents a deterministic slope stability analysis based on strength reduction finite-element method first. After that, three simple probabilistic methods, including FOSM, FORM, and Monte Carlo method, are introduced to perform a simple probabilistic slope stability analysis. Finally, the RFEM approach combining random field generation techniques and finite-element methods is shown and applied to slope stability risk assessment.

The elastoplastic finite-element slope stability method makes no a priori assumptions about the shape or location of the critical failure mechanism, offering significant benefits over traditional limit equilibrium methods on slope stability analysis.

FOSM, FORM, and Monte Carlo method are relatively basic and practical probabilistic analysis methods. Based on different algorithms, the uncertainty and randomness of the soil properties, especially the mean and standard deviation, can be taken into account from different views. However, there are some deficiencies, such as limit of accuracy and time-consuming on these methods.

The RFEM approach combines finite-element slope stability method and local averaging subdivision method, which can take full account of spatial correlation and local averaging. The influence of spatial correlation length and coefficient of variance on the probability of failure can be studied using a parametric approach. In the elastoplastic RFEM, the failure mechanism is free to seek out the weakest path through the soil, which leads to higher probabilities of failure than that conducted by finite-element local averaging alone.

In summary, simplified probabilistic analysis in which spatial variability is ignored can lead to unconservative estimates of the probability of failure, while the RFEM approach that considers spatial correlation and local averaging would be a practical method on slope stability risk assessment.

Acknowledgements

The author wishes to acknowledge the support from Professor D.V. Griffiths for his supervision during the period of the author's visiting scholar at Colorado School of Mines. The author also acknowledges the support of the National Key Research and Development Program of China Grant No. 2018YFC1508602 and National Natural Science Foundation of China Grant No. 51539006.

Author details

Yijiang Zhang*, Enyue Ji and Weiwei Xu
Geotechnical Engineering Department, Nanjing Hydraulic Research Institute (NHRI), Nanjing, China

*Address all correspondence to: zhangyijiang1025@163.com

References

[1] Duncan J, Wright S. The accuracy of equilibrium methods of slope stability analysis. Engineering Geology. 1980;**16** (1–2):5-17. DOI: 10.1016/0013-7952(80) 90003-4

[2] Deng D, Zhao L, Li L. Limit equilibrium slope stability analysis using the nonlinear strength failure criterion. Canadian Geotechnical Journal. 2015;**52**(5):563-576. DOI: 10.1139/cgj-2014-0111

[3] Liu S, Shao L, Li H. Slope stability analysis using the limit equilibrium method and two finite element methods. Computers and Geotechnics. 2015;**63**:291-298. DOI: 10.1016/j.compgeo.2014.10.008

[4] Tschuchnigg F, Schweiger H, Sloan S. Slope stability analysis by means of finite element limit analysis and finite element strength reduction techniques. Part I: Numerical studies considering non-associated plasticity. Computers and Geotechnics. New Jersey, USA: John Wiley & Sons, Inc.; 2015;**70**:169-177. DOI: 10.1016/j.compgeo.2015.06.018

[5] Griffiths D, Lane P. Slope stability analysis by finite elements. Géotechnique. 1999;**49**(3):387-403. DOI: 10.1680/geot.1999.49.3.387

[6] Smith I, Griffiths D, Margetts L. Programming the Finite Element Method. 5th ed. Hoboken, New Jersey, USA: John Wiley & Sons, Ltd; 2013. DOI: 10.1002/9781119189237

[7] Christian J, Ladd C, Baecher G. Reliability applied to slope stability analysis. Journal of Geotechnical Engineering. 1994;**120**(12):2180-2207. DOI: 10.1061/(ASCE)0733-9410

[8] Whitman R. Organizing and evaluating uncertainty in geotechnical engineering. Journal of Geotechnical Engineering. 2000;**126**(7):583-593. DOI: 10.1061/(ASCE)1090-0241

[9] Duncan J. Factors of safety and reliability in geotechnical engineering. Journal of Geotechnical Engineering. 2000;**126**(4):307-316. DOI: 10.1061/(ASCE)1090-0241

[10] El-Ramly H, Morgenstern N, Cruden D. Probabilistic slope stability analysis for practice. Canadian Geotechnical Journal. 2002;**39**(3): 665-683. DOI: 10.1139/t02-034

[11] Cho S. Probabilistic assessment of slope stability that considers the spatial variability of soil properties. Journal of Geotechnical and Geoenvironmental Engineering. 2010;**136**(7):975-984. DOI: 10.1061/(ASCE)GT.1943-5606.0000309

[12] Liu X, Li D, Cao Z, Wang Y. Adaptive Monte Carlo simulation method for system reliability analysis of slope stability based on limit equilibrium methods. Engineering Geology. 2020;**264**:105384. DOI: 10.1016/j.enggeo.2019.105384

[13] Griffiths D, Fenton G. Probabilistic slope stability analysis by finite elements. Journal of Geotechnical and Geoenvironmental Engineering. 2004; **130**(5):507-518. DOI: 10.1061/(ASCE) 1090-0241

[14] Griffiths D, Marquez R. Three-dimensional slope stability analysis by elasto-plastic finite elements. Géotechnique. 2007;**57**(6):537-546. DOI: 10.1680/geot.2007.57.6.537

[15] Zhu D, Griffiths D, Fenton G. Worst-case spatial correlation length in probabilistic slope stability analysis. Géotechnique. 2019;**69**(1):85-88. DOI: 10.1680/jgeot.17.T.050

[16] Griffiths D, Fenton G. Risk Assessment in Geotechnical Engineering. John Wiley & Sons, Inc; 2008

[17] Li K, Lumb P. Probabilistic design of slopes. Canadian Geotechnical Journal. 1987;**24**(4):520-535. DOI: 10.1139/t87-068

[18] Abdallah I, Malkawi H, Hassan W, Abdulla F. Uncertainty and reliability analysis applied to slope stability. Structural Safety. 2000;**22**(2):161-187. DOI: 10.1016/S0167-4730(00)00006-0

[19] Suchomel R, Mašín D. Comparison of different probabilistic methods for predicting stability of a slope in spatially variable c–ϕ soil. Computers and Geotechnics. 2010;**37**:132-140. DOI: 10.1016/j.compgeo.2009.08.005

[20] Wu Z, Li Y, Chen J, Zhang H, Pei L. A reliability-based approach to evaluating the stability of high rockfill dams using a nonlinear shear strength criterion. Computers and Geotechnics. 2013;**51**:42-49. DOI: 10.1016/j.compgeo.2013.01.005

[21] Wu Z, Shi Q, Guo Q, Chen J. CST-based first order second moment method for probabilistic slope stability analysis. Computers and Geotechnics. 2017;**85**:51-58. DOI: 10.1016/j.compgeo.2016.12.017

[22] Hassan A, Wolff T. Search algorithm for minimum reliability index of earth slopes. Journal of Geotechnical and Geoenvironmental Engineering. 1999;**125**(4):301-308. DOI: 10.1061/(ASCE)1090-0241(1999)125:4(301)

[23] Hasofer A, Lind N. Exact and invariant second-moment code format. Journal of the Engineering Mechanics Division. 1974;**100**:111-121

[24] Abramson L, Lee T, Sharma S, Boyce G. Slope Stability and Stabilization Methods. 2nd ed. John Wiley & Sons, Inc; 2002

[25] Low B, Tang W. Efficient spreadsheet algorithm for first-order reliability method. Journal of Engineering Mechanics. 2007;**133**(12):1378-1387. DOI: 10.1061/(ASCE)0733-9399(2007)133:12(1378)

[26] Low B. FORM, SORM, and spatial modeling in geotechnical engineering. Structural Safety. 2014;**49**:56-64. DOI: 10.1016/j.strusafe.2013.08.008

[27] Tobutt D. Monte Carlo simulation methods for slope stability. Computers and Geosciences. 1982;**8**(2):199-208. DOI: 10.1016/0098-3004(82)90021-8

[28] Jiang S, Li D, Cao Z, Zhou C, Phoon K. Efficient system reliability analysis of slope stability in spatially variable soils using Monte Carlo simulation. Journal of Geotechnical and Geoenvironmental Engineering. 2015;**141**(2):04014096. DOI: 10.1061/(ASCE)GT.1943-5606.0001227

[29] Aladejare A, Akeju V. Design and sensitivity analysis of rock slope using Monte Carlo simulation. Geotechnical and Geological Engineering. 2020;**38**:573-585. DOI: 10.1007/s10706-019-01048-z

[30] Fenton G, Vanmarcke E. Simulation of random fields via local average subdivision. Journal of Engineering Mechanics. 1990;**116**(8):1733-1749. DOI: 10.1061/(ASCE)0733-9399(1990)116:8(1733)

Chapter 8

Earth-Rock Dams' Breach Modelling

Qiming Zhong, Yibo Shan and Jiaxin Liu

Abstract

Simulation of dam breach process has significant influence on the evaluation of consequence of dam breach flood. In this study, research progresses on the numerical modeling of earth-rock dams' breach process are summarized, especially the latest research results of the author's research team in recent years. However, there still has a considerable gap in the versatility of computer software and visualization technology of dam breaching process. It is suggested that more efforts should be made in the future to study the detailed physically based numerical model for core dam and concrete face rockfill dam; further, more attention should be paid to the application of visualization technology in dam breach process simulation. Finally, the universal and friendly visualization computer software that can accurately simulate the dam failure process and flood routing for earth-rock dams is sorely needed.

Keywords: earth-rock dam, numerical model, computer software, research progress

1. Introduction

China has nearly 100,000 reservoir dams, of which earth-rock dams account for more than 95% [1, 2]. Most of these reservoir dams were built in the 1950s and 1970s. Due to economic and technical conditions at that time, the problem of dangerous reservoirs in China was outstanding [3]. According to statistics [4], from 1954 to 2018, 3541 reservoir dams broke in China. The "75·8" flood occurred in Henan in 1975, which led to the collapse of 2 large reservoirs in Banqiao and Shimantan (**Figures 1** and **2**), 2 medium-sized reservoirs in Tiangang and Zhugou, and 58 small reservoirs, causing heavy casualties and property losses [5]. In the twenty-first century, with the improvement of the dam safety management level and the comprehensive development of the reservoir's risk elimination and reinforcement, the number of dam breaks has been significantly reduced, but due to the frequent occurrence of extreme weather events, dams' breaching still occur frequently. On July 19, 2018, the Zenglongchang Reservoir in Inner Mongolia and the Sheyuegou Reservoir in Xinjiang on August 1, 2018, successively dams' breaching [4] (**Figures 3** and **4**).

Therefore, it is necessary to establish a mathematical model and numerical calculation method that reasonably simulates the process of overtopping and seepage failure collapse, improves the prediction accuracy of the flood flow process of earth-rock dam collapse, and provides theoretical and technical support for the

Figure 1.
Final breach of Banqiao dam.

Figure 2.
Final breach of Shimantan dam.

Figure 3.
Final breach of Zenglongchang dam.

Figure 4.
Final breach of Sheyuegou dam.

evaluation of the consequences of dam collapse and the preparation of emergency plans. This article will briefly introduce the research progresses on the mechanisms and numerical models of earth-rock dams' breaching, especially the latest research results of the author's research team in recent years, and make suggestions for future research.

2. Study on mathematical model of earth-rock dam break

The mathematical model of earth-rock dams' breaching is generally divided into three categories [6]: The first category is the parameter model. Most of these models are based on statistical analysis of dam-break case data, and empirical formulas are used to calculate and obtain dam-break-related parameters. Although most models cannot consider the erosion characteristics of damming materials, but the parameter model formula is simple and fast to calculate and is also often used for rapid evaluation of the consequences of dams' breaching. The second category is a simplified mathematical model based on the mechanism of failure. It is generally assumed that the shape of the fractured breach (rectangular, inverted trapezoidal, triangular, etc.) remains unchanged during the dams' breaching. The method based on the flow shear stress and the critical shear stress of the dam material or the erosion formula of the dam material is used to calculate the breach development process; the weir flow is used (overtopping dam failure) or pore flow (seepage failure dam breaching) formulas are used to calculate the breach flow. The stability analysis of the breach slope mostly uses the limit equilibrium method; generally, the numerical calculation method based on time step iteration is used to simulate the breach development process and the breach flow process. The advantage of this type of model is that it considers the failure mechanism of earth-rock dams, and the calculation speed is relatively fast, which is the most widely used in the numerical simulation of earth-rock dams dam breaching process. The third category is a detailed mathematical model based on the failure mechanism. In recent years, a series of researches on one-dimensional, average two-dimensional, and three-dimensional mathematical models based on the hydrodynamic dam material erosion equation have made significant progress, which can simulate the dams' breaching process of earth-rock dams in more detail. In order to deal with the diffuse overtopping flow composed of discontinuous mixed flow states, shock wave capturing methods such as approximate Riemann solution method and total variation declining (TVD) method are generally used, and finite volume method, level set method, and smooth particle hydrodynamic method are used to solve the governing equation. This type of model is a fast-developing simulation method in recent years, but it can only be used for the simulation of the overtopping collapse process of homogeneous dams or landslide dams. It has not been used to simulate the process of seepage and failure of earth-rock dams and the simulation of the process of overtopping failure of other types of earth-rock dams [6].

2.1 Parametric model

In 1977, Kirkpatrick [7] proposed the first empirical formula for predicting peak outflow Q_p, and then scholars from various countries proposed a series of models. With the continuous enrichment of dam failure case investigation data and the deepening of research, the dam failure parameter model has gradually evolved from the single-parameter model to a multi-parameter model, and the output results have increased from the original peak outflow of the breach to the final average width of the breach and the duration of the dam and can consider the shape of the dam body, reservoir capacity, dam material characteristics, etc. The peak outflow rate of breach is very important for the evaluation of the consequences of dam breaching. Therefore, domestic and foreign scholars have studied more. The commonly used parameter model of peak outflow rate is shown in **Table 1**.

In 1988, the US Bureau of Reclamation (USBR) [14] proposed the first empirical formula for predicting the final average width of the breach B_{ave}, and then scholars from various countries put forward a series of models. The commonly used parameter model of the final average width of the breach is shown in **Table 2**.

In 1984, MacDonald and Langridge-Monopolis [42] proposed the first empirical formula for predicting the duration of dam failure, and then scholars from various countries proposed a series of models. Commonly used dam-break duration parameter model is shown in **Table 3**.

Due to the difficulties in obtaining the dam-break duration, the relatively low accuracy of the data, and the small number of samples, the dam-break duration model has a large deviation in the calculation of individual cases.

In order to fully consider the dam type, dam breach mode, reservoir characteristics, and breach characteristics, the reservoir capacity (V_w) is above the bottom of the breach at the dam break, the water depth (h_w) above the bottom of the dam at the dam break (h_d), and the final depth of the rupture (h_b). For other parameters, the method of statistical regression is used to obtain the results of the peak flow of the breach, the final average width of the breach, and the duration of the dam breach. From the above statistics, it can be seen that the parameter model can simulate the dam-break parameters simply and quickly, which is an efficient and rapid evaluation method, but the parameter model cannot provide the dam-break flood flow process line.

2.2 Simplified mathematical model based on failure mechanism

In the 1960s, European and American scholars began to study a simplified mathematical model based on the mechanism of collapse based on hydraulics and sediment transport formulas. This model is also the most widely used mathematical model of earth-rock dams' breaching. In 1965, from the US Bureau of Reclamation, Cristofano [27] established the first mathematical model of homogeneous dam overtopping failure. Afterward, scholars from various countries proposed a series of mathematical models for simulating earth-rock dam collapse [6, 28]. The most widely used is the NWS BREACH model developed by Fread from the National Weather Service [29]. In recent years, the Nanjing Hydraulic Research Institute and China Institute of Water Resources and Hydropower Research have conducted

Model	Case number	Expression
Kirkpatrick (1977) [7]	19	$Q_p = 1.268(h_w + 0.3)^{2.5}$
Soil Conservation Service (1981) [8]	13	$Q_p = 16.6{h_w}^{1.85}$
Hagen (1982) [9]	6	$Q_p = 0.54(h_d S)^{0.5}$
Singh and Snorrason (1984) [10]	28	$Q_p = 13.4{h_d}^{1.89}$ or $Q_p = 1.776S^{0.47}$
MacDonald and Langridge-Monopolis (1984) [11]	23	$Q_p = 1.154(V_w h_w)^{0.412}$
Costa (1985) [12]	31	$Q_p = 0.981(h_d S)^{0.42}$
Evans (1986) [13]	29	$Q_p = 0.72{V_w}^{0.53}$
USBR (1988) [14]	21	$Q_p = 19.1{h_w}^{1.85}$
Froehlich (1995) [15]	22	$Q_p = 0.607{V_w}^{0.295}{h_w}^{1.24}$
Walder and O'Connor (1997) [16]	18	$Q_p = 0.031\,g^{0.5}{V_w}^{0.47}{h_w}^{0.15}{h_b}^{0.94}$
Xu and Zhang[1] (2009) [17]	75	$Q_p = 0.175\,g^{0.5}{V_w}^{5/6}(h_d/h_r)^{0.199}({V_w}^{1/3}/h_w)^{-1.274}e^{B}{}_4$
Pierce et al. (2010) [18]	87	$Q_p = 0.0176(Vh)^{0.606}$or $Q_p = 0.038\ V^{0.475}\ h^{1.09}$
Thornton et al. (2011) [19]	38	$Q_p = 0.1202\ L^{1.7856}$ or $Q_p = 0.863\ V^{0.335}{h_d}^{1.833}{W_{ave}}^{-0.663\text{or}}Q_p = 0.012\ V^{0.493}{h_d}^{1.205}L^{0.226}$
Lorenzo and Macchione (2014) [20]	14	$Q_p = 0.321\,g^{0.258}(0.07V_w)^{0.485}{h_b}^{0.802}$(overtopping) $Q_p = 0.347\,g^{0.263}(0.07V_w)^{0.474}{h_b}^{-2.151}{h_w}^{2.992}$(seepage failure)
Hooshyaripor et al. (2014) [21]	93	$Q_p = 0.0212\ V^{0.5429}h^{0.8713}$ or $Q_p = 0.0454\ V^{0.448}h^{1.156}$
Azimi et al. (2015) [22]	70	$Q_p = 0.0166(gV)^{0.5}h$
Froehlich[2] (2016) [23]	41	$Q_p = 0.0175k_M k_H(gV_w h_w {h_b}^2/W_{ave})^{0.5}$
Mei Shiang et al. (2018) [24]	154	$Q_p = V_u g^{0.5}{h_w}^{-0.5}({V_w}^{1/3}/h_w)^{-1.58}(h_w/h_b)^{-0.76}(h_d/h_0)^{0.10}\mathrm{e}^{-4.55}$(homogeneous dam) $Q_p = V_u g^{0.5}{h_w}^{-0.5}({V_w}^{1/3}/h_w)^{-1.51}(h_w/h_b)^{-1.09}(h_d/h_0)^{-0.12}\mathrm{e}^{-3.61}$ (core-wall dam)

Q_p *is the peak outflow of the breach;* h_w *is the water depth above the bottom of the breach when the dam breaks;* h_d *is the height of the dam;* S *is the reservoir capacity;* V_w *is the reservoir capacity above the bottom of the breach when the dam breaks;* g *is the gravity acceleration;* h_b *is the depth of the dam breaks;* h_r *is the reference dam height, take 15 m;* V *is the reservoir capacity at dam breaching;* h *is the water level at dam breaching;* L *is the length of the dam;* W_{ave} *is the average width of the dam;* k_M *and* k_H *are coefficients.*
[1]*The expression of parameter* B_4 *is* $B_4 = b_3 + b_4 + b_5$*, for core-wall dam, concrete face rockfill dam or homogeneous dam,* b_3 *is taken as −0.503, 0.591, or − 0.649, respectively; for overtopping or seepage failure,* b_4 *is taken as −0.705 or − 1.039, respectively; for dam materials with high, medium, or low erosion rate,* b_5 *is taken as −0.007, −0.375, or − 1.362, respectively.*
[2]*For overtopping dam failure,* $k_M = 1.85$*; for seepage failure dam failure,* $k_M = 1$*; when* $h_b \le 6.1\ m, k_H = 1$*; when* $h_b > 6.1\ m$*,* $k_H = (h_b/6.1)^{1/8}$*.*

Table 1.
Parameter model of peak outflow rate.

systematic research work on the mathematical model of earth-rock dams' breaching, establishing NHRI-DB series and DB-IWHR series dam-break mathematical models, respectively. The commonly used simplified mathematical model of earth-rock dam breaching is shown in **Table 4**.

It can be seen from the above analysis that this type of model is mainly aimed at the two failure modes of earth-rock dam overtopping and seepage failure. By

Model	Case number	Expression
USBR (1988) [14]	21	$B_{ave} = 3h_w$
Von Thun and Gillette[1] (1990) [25]	57	$B_{ave} = 2.5\,h_w + C_b$
Froehlich[2] (1995) [26]	22	$B_{ave} = 0.1803\,K_0(V_w)^{0.32}(h_b)^{0.19}$
Xu and Zhang[3] (2009) [17]	75	$B_{ave} = 0.787(h_b)(h_d/h_r)^{0.133}(V_w^{1/3}/h_w)^{0.652}e^{B_3}$
Froehlich[4] (2016) [23]	41	$B_{ave} = 0.27k_M(V_w)^{1/3}$
Mei Shiang et al. (2018) [24]	154	$B_{ave} = h_b(V_w^{1/3}/h_w)^{0.84}(h_w/h_b)^{2.30}(h_d/h_0)^{0.06}e^{-0.90}$(homogeneous dam) $B_{ave} = h_b(V_w^{1/3}/h_w)^{0.55}(h_w/h_b)^{1.97}(h_d/h_0)^{-0.07}e^{-0.09}$(core-wall dam)

[1]When $S < 1.2335 \times 10^6\ m^3$, $C_b = 6.096$; when $1.2335 \times 10^6\ m^3 \le S < 6.1676 \times 10^6\ m^3$, $C_b = 18.288$; when $6.1676 \times 10^6\ m^3 \le S < 1.2335 \times 10^7\ m^3$, $C_b = 42.672$; when $S \ge 1.2335 \times 10^7\ m^3$, $C_b = 54.864$.
[2]For overtopping dam failure, $K_0 = 1.4$; for seepage failure dam breaching, $K_0 = 1.0$.
[3]h_r is the dam height, which is 15 m; the expression of parameter B_3 is $B_3 = b_3 + b_4 + b_5$, for core-wall dam, concrete face rockfill dam or homogeneous dam, b_3 takes −0.041, 0.026 or 0.226; for overtopping or seepage failure, $b_4 = 0.149$ or − 0.389, respectively; for dams with high, medium, or low erosion rate, b_5 is 0.291, 0.14, or 0.391, respectively.
[4]For overtopping dam failure, $k_M = 1.3$; for seepage failure dam breaching, $k_M = 1.0$.

Table 2.
Parameter model of the final average width of the breach.

Model	Case number	Expression
MacDonald and Langridge-Monopolis (1984) [11]	23	$T_f = 0.0179(0.0261(V_w h_w)^{0.769})^{0.364}$
USBR (1988) [14]	21	$T_f = 0.011B_{ave}$
Froehlich (1995) [26]	22	$T_f = 0.00254(V_w)^{0.53}(h_b)^{-0.9}$
Xu and Zhang[1] (2009) [17]	75	$T_f = 0.304T_r(h_d/h_r)^{0.707}(V_w^{1/3}/h_w)^{1.228}e^{B_5}$
Froehlich (2016) [23]	41	$T_f = 63.2(V_w/(gh_b^2))^{0.5}$
Mei Shiang et al.[2] (2018) [24]	154	$T_f = T_0(V_w^{1/3}/h_w)^{0.56}(h_w/h_b)^{-0.85}(h_d/h_0)^{-0.32}e^{-0.20}$(homogeneous dam) $T_f = T_0(V_w^{1/3}/h_w)^{1.52}(h_w/h_b)^{-11.36}(h_d/h_0)^{-0.43}e^{-1.57}$(core-wall dam)

[1]Tr means the duration of the reference dam break, take 1 h; the expression of parameter B5 is B5 = b3 + b4 + b5, for core-wall dam, concrete face rockfill dam, or homogeneous dam, b3 takes −0.327, −0.674, or − 0.189; for overtopping or seepage failure, b4 = −0.579 or − 0.611, respectively; for dam materials with high, medium, or low erosion rate, b5 is −1.205, −0.564, or 0.579 respectively.
[2]T_0 means unit duration, take 1 h.

Table 3.
Dam-break duration parameter model.

assuming the shape of the breach, different flow calculation formulas and erosion formulas are used to simulate the scouring of the dam material, and different simulation methods are used to analyze the vertical undercut and lateral expansion of the breach. Most of the models use iterative numerical calculation methods based on time steps to simulate the process of dam break and can output the parameters of dam break (such as the flow of the breach, the size of the breach, the water level of the reservoir, etc.) at each time step.

For example, based on the overtopping breach mechanism of the clay-core wall dam, a mathematical model to simulate its breach process is proposed. The model is based on the shape of the dam body and the characteristics of the flood flow to

Model	Shape of breach	The flow of the breach	Erosion formula	Mechanical analysis	Breach mode	Type of dam
Cristofano (1965) [27]	Trapezoid	Wide crest weir formula	Cristofano formula	Breach without lateral collapse	Overtopping	Homogeneous
BRDAM (1981) [30]	Parabolic	Wide crest weir formula, vent flow formula	Schoklitsch formula	Breach without lateral collapse (overtopping), top collapse (seepage)	Overtopping or seepage	Homogeneous
DAMBRK (1984) [31]	Trapezoid or rectangle	Wide crest weir formula	Even flush	Breach without lateral collapse	Overtopping	Homogeneous
BEED (1985) [32]	Trapezoid	Wide crest weir formula	Einstein and Brown formula, Meyer-Peter-Mueller formula	Collapse laterally	Overtopping	Homogeneous
NWS BREACH (1988) [29]	Trapezoid or rectangle	Wide crest weir formula, vent flow formula	Correction Meyer-Peter-Mueller formula	Collapse laterally (overtopping), top collapse (seepage)	Overtopping or seepage	Homogeneous, core wall
HR BREACH (2002, 2009) [33, 34]	Effective stress method	1D stable non-uniform weir flow formula	Sediment transport formula or erosion rate formula	Single (two) side erosion, collapse laterally, stability analysis of core wall	Overtopping or seepage	Homogeneous earth dam, core wall
FIREBIRD (2006) [35]	Trapezoid	Unsteady Saint-Venant equation	Sediment transport formula or erosion rate formula	Collapse laterally	Overtopping	Homogeneous
WinDAM/SIMBA (2005, 2006, 2010) [36–38]	Rectangle	Wide crest weir formula	Erosion rate formula	Breach without lateral collapse	Overtopping	Homogeneous
DLBreach (2013) [39]	Trapezoid	Wide crest weir formula, vent flow formula	Sediment transport formula or erosion rate formula	Single (two) side erosion, collapse laterally, stability analysis of core wall, dam foundation erosion	Overtopping or seepage	Homogeneous, core wall
Hong Kong University of Science and Technology model [40]	Trapezoid	Wide crest weir formula, vent flow formula	Erosion rate formula	Collapse laterally (overtopping), top collapse (seepage)	Overtopping or seepage	Homogeneous, landslide

Model	Shape of breach	The flow of the breach	Erosion formula	Mechanical analysis	Breach mode	Type of dam
DB-IWHR series dam-break mathematical model of China Institute of Water Resources and Hydropower Research [41–43]	Trapezoid	Wide crest weir formula	Erosion rate formula	Collapse laterally	Overtopping	Homogeneous, core wall, landslide dam
NHRI-DB series dam-break mathematical model of Nanjing Hydraulic Research Institute [44–48]	Trapezoid	Wide crest weir formula, vent flow formula	Sediment transport formula or erosion rate formula	Shearing or dumping of the core wall, panel break, collapse laterally (overtopping), top collapse (seepage)	Overtopping or seepage	Homogeneous, core wall, face dam, landslide

Table 4.
Simplified mathematical model of earth-rock dam breaching.

determine the initial scoring position of the downstream slope during erosion. The flow formula of the wide crested weir is used to calculate the rupture flow. The mechanical equilibrium method is used to simulate the tipping and shear failure of the core wall; the model can also consider the erosion of the dam body on one side, the erosion on both sides, and the erosion of the dam foundation and the process of water and soil coupling during dam break.

Based on the mechanism revealed by the model test of the overtopping breach of the homogeneous cohesive earth dam, the author has established a mathematical model that can simulate its collapse process (**Figure 5**). The specific modules of the model are as follows.

This model is based on the shape of the dam body and the characteristics of the flow at the top of the crater to determine the formation position of the "dark ridge." The traceable erosion formula that can consider the physical and mechanical characteristics of the dam material is used to simulate the movement of the "dark ridge." The collapse of the dam body: choose a reasonable erosion formula of the dam material to simulate the development of the dam crest and the downstream slope failure, and use the limit equilibrium method to simulate the failure of the collapse slope. The model considers incomplete dam failure and erosion of the dam foundation, as well as erosion on one side and both sides of the dam body.

The flow chart of the model calculation process of the collapse process of the homogeneous earth dam is shown in **Figure 6**.

There are two major highlights of the NHRI-DB concrete-face dam-break mathematical model [47]: the adoption of total-load nonequilibrium transport equation (Eq. (1)) [49] to simulate the erosion process of sand gravels with a wide range of gradation and the establishment of an analogy to simulate the failure process of each concrete-face slab under various loads during the dam breaching process.

$$\frac{\partial(AC_t)}{\partial t}+\frac{\partial(Q_bC_t)}{\partial x}=-\frac{Q_b}{L_S}(C_t-C_{t*}) \tag{1}$$

where t = time; x = longitudinal coordinate; A = cross-sectional flow area in the breach channel; C_t = actual total-load sediment concentration; C_t = sediment concentration at the equilibrium state; and L_s = adaptation length characterizing the adjustment of sediment from a nonequilibrium state to equilibrium state.

In the NHRI-DB core dam-break mathematical model [45], a hydraulic method was used to predict the initial scour position for high dam. A time averaged erosion equation was adopted to simulate the backward erosion of dam's shoulder.

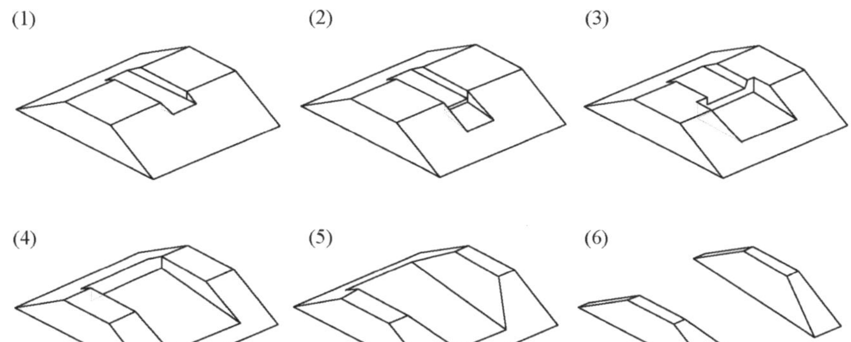

Figure 5.
Schematic diagram of the author's model calculation process. (1) Breach formation, (2) Scarp formation, (3) Scarp widen, (4) Headcut scour, (5) Breach widen, (6) Breach fully formed.

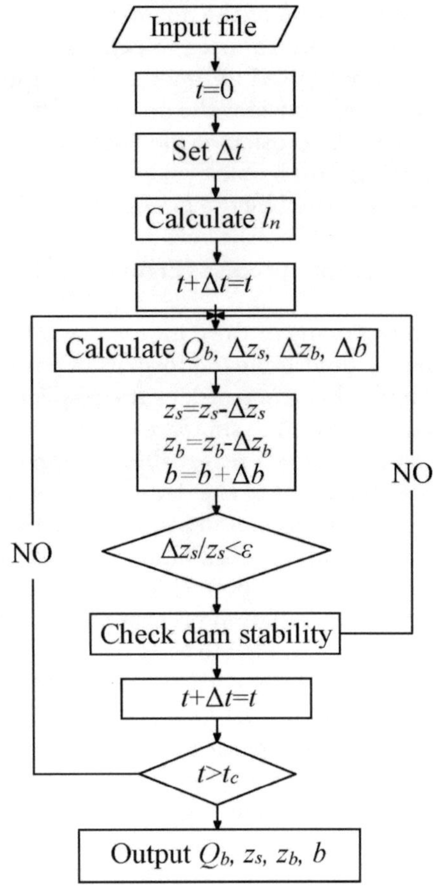

Figure 6.
Calculation flow chart of the process of overburden collapse of the homogeneous earth dam.

The broad-crested weir equation (Eqs. (2) and (3)) [50, 51] was adopted to calculate the breach flow discharge. Furthermore, the sliding or overturning failure was adopted as the key mechanism for the core, which was judged based upon numerical analysis. The calculated results show that the proposed model gives reasonable peak outflow, final breach width, and failure time.

$$Q_b = k_{sm}\left(c_1 B_b H^{1.5} + c_2 m H^{2.5}\right) \tag{2}$$

where B_b is the bottom width of the breach (m), H represents the difference "$z_s - z_b$" (m), in which z_b is the elevation of the breach bottom (m), m is the side slope (horizontal/vertical) of the breach, c_1 and c_2 are the discharge coefficients with values of 1.7 m0.5/s and 1.3 m0.5/s [50], and k_{sm} is the submergence correction factor for tailwater effects on weir outflow.

$$k_{sm} = \begin{cases} 1.0 & \dfrac{z_t - z_b}{z_s - z_b} < 0.67 \\ 1.0 - 27.8\left(\frac{z_t - z_b}{z_s - z_b} - 0.67\right)^3 & \text{otherwise} \end{cases} \tag{3}$$

where z_t is the tailwater level (m).

The advantage of this type of model is that it can consider the failure mechanism of the earth-rock dam and can use a short calculation time to complete the

Model	Determination method of breach shape	Flow of breach	Dam material erosion	Mechanical analysis	Calculation method
Wang and Bowles (2006) [53]	Scour without sediment motion	Shallow water equations	Erosion rate formula	Three-dimensional collapse laterally	Finite different method
Faeh (2007) [54]	Two-dimensional Exner equations	Shallow water equations	Traction load and suspended load formula	Collapse laterally	Finite volume method
Wu et al. (2007, 2012) [55, 56]	One- and two-dimensional nonequilibrium total sand transport equations	General shallow water equations	Total sand transport formula	Collapse laterally	Finite volume method
Swartenbroekx et al. (2010) [57]	Two-dimensional Exner equations	Shallow water equations	Traction load formula	Collapse laterally	Finite volume method
Li et al. (2011) [58]	Two-dimensional nonequilibrium sediment transport equations (suspended load)	Shallow water equations	Empirical formulas of sediment carrying rate and Sedimentation rate	Breach without lateral collapse	Finite volume method
Cao et al. (2011) [59]	Two-dimensional nonequilibrium total sediment transport equations	General shallow water equations	Traction load formula	Collapse laterally	Finite volume method
Rosatti and Begnudelli (2013) [60, 61]	Two-dimensional mass conservation and energy conservation equations (solid phases)	Shallow water equations (liquid)	Floe concentration formula	Breach without lateral collapse	Finite volume method
Juez et al. (2013, 2014) [62, 63]	One- and two-dimensional Exner equations	Saint-Venant equations and shallow water equations	10 different erosion formulas	Breach without lateral collapse	Finite volume method
Swartenbroekx et al. (2013) [64]	Two-dimensional mass conservation and energy conservation equations (traction load)	Shallow water equations (clean water)	Erosion rate formula	Breach without lateral collapse	Finite volume method
Guan et al. (2014) [65]	Two-dimensional nonequilibrium sediment transport equations (traction load)	Shallow water equations (pure water)	Traction load formula	Collapse laterally	Finite volume method
Kesserwani et al. (2014) [66]	Two-dimensional nonequilibrium sediment transport equations (suspended load)	Shallow water equations	Empirical formulas of sediment carrying rate and	Breach without lateral collapse	Intermittent Galerkin method

Model	Determination method of breach shape	Flow of breach	Dam material erosion	Mechanical analysis	Calculation method
			sedimentation rate		
Razavitoosi et al. (2014) [67]	N-S equations (solid phases, non-Newtonian fluid)	N-S equations (liquid, non-Newtonian fluid)	/	Breach without lateral collapse	Smoothed particle hydrodynamics method
Marsooli and Wu (2015) [68]	Three-dimensional nonequilibrium sediment transport equations	N-S equations	Traction load and suspended load formula	Breach without lateral collapse	Finite volume method and volume of fluid
Abderrezza et al. (2016) [69]	Two-dimensional Exner equations	Shallow water equations	Traction load formula	Collapse laterally	Finite volume method
Cantero-Chinchilla et al. (2016) [70]	One-dimensional nonequilibrium sediment transport equations	Saint-Venant equations, vertical momentum equation	Traction load and suspended load formula	Breach without lateral collapse	Finite volume method
Cristo et al. (2016, 2018) [71, 72]	Two-dimensional mass conservation and energy conservation equations (solid phase)	Shallow water equations (liquid)	Traction load formula	Bed collapse algorithm	Finite volume method
YAN Zhikun et al. (2019) [73]	Two-dimensional nonequilibrium sediment transport equations	General shallow water equations	Total sand transport formula considering bed slope	Bed collapse algorithm	Finite volume method

Table 5.
Detailed mathematical model of earth-rock dams' breaching.

simulation of the dam-break process; however, most models cannot really consider the water-soil coupling effect during the dam-break process.

2.3 Detailed mathematical model based on failure mechanism

In order to fully describe the water-soil coupling effect in the process of dams' breaching, in recent years, with the improvement of computer performance and the development of sediment science and computational fluid dynamics, a series of nonequilibrium dam material transport theory has emerged based on shallow water hypothetical detailed mathematical model for dam failure [52]. The commonly used detailed mathematical model of earth-rock dams' breaching is shown in **Table 5**.

It can be seen from the above statistics that this type of model is mainly based on the continuity equations of water flow (Eq. (4)), momentum equations (Eq. (5)), and energy equations (Eq. (6)), coupled with the sediment movement equation, and the finite volume method and other numerical simulation methods are used to discretely solve the governing equations

$$\frac{\partial \rho}{\partial t} + \frac{\partial(\rho u_x)}{\partial x} + \frac{\partial(\rho u_y)}{\partial y} + \frac{\partial(\rho u_z)}{\partial z} = 0 \tag{4}$$

$$\frac{\partial \vec{v}}{\partial t} + \left(\vec{v} \cdot \nabla\right)\vec{v} = -\frac{1}{\rho}\nabla p + \vec{f} + \frac{1}{\rho}\overrightarrow{F_v} \tag{5}$$

$$\frac{\partial}{\partial t}\left[\rho\left(e + \frac{v^2}{2}\right)\right] + \nabla \cdot \left[\rho\left(e + \frac{v^2}{2}\right)\vec{v}\right] = \rho\dot{q} - \nabla \cdot \left(p\vec{v}\right) + \rho\left(\vec{f} \cdot \vec{v}\right) + \dot{Q}_v + \dot{W}_v \tag{6}$$

In Yan Zhikun's model [73], based on the continuity equations of water flow, momentum equations, and nonequilibrium sediment transport equations, a planar two-dimensional mathematical model of dam rupture along the depth average is proposed. The sand capacity and the collapse mechanism of the two-dimensional slope during the dam-break process. The fully coupled method is used to convert the hydrodynamic equation and the nonequilibrium sediment transport equation into a shallow water equation with source terms and is based on the finite volume method under a rectangular grid. Discrete processing, using conservative, non-negative water depth numerical reconstruction format to make the model have second-order accuracy in the space–time direction, using HLLC [74] approximate Riemann solver to calculate grid boundary flux, SGM (Surface Gradient Method) format to calculate water surface gradient source terms, semi-implicit format. For the bottom bed friction term, the explicit gradient calculation of the source term of the concentration gradient is used to numerically solve the control equation.

Such models can achieve detailed simulation of the dam-break process, but the calculation speed is slow, and it can only be used for the numerical simulation of the overtopping dams' breaching. However, this method can fully consider the coupling effect of water-soil coupling in the process of dam failure and can simulate complex boundary conditions, which is the development direction of numerical simulation of earth-rock dams' breaching.

3. Conclusion and suggestions

Earth-rock dams' breaching mechanism and dam-break process simulation are the foundation of dam-break disaster assessment and emergency response. They involve fluid mechanics, sediment kinematics, soil mechanics, and other disciplines. They are complex water-soil coupling problems. After decades of research and exploration, various mathematical models of dams' breaching have been developed and made a series of innovative achievements, which provide theoretical support for improving the accuracy of flood disaster prediction of earth-rock dams. It is suggested that in the future, research efforts should be intensified on the mathematical model of the detailed simulation of the earth-rock dam breaching process, focusing on the application of visualization technology in the simulation of the dam-break process and accelerating the development of a universal and friendly simulation of the earth-rock dams' breaching and the visual calculation of the disaster-causing process software.

Author details

Qiming Zhong[1]*, Yibo Shan[1] and Jiaxin Liu[2]

1 Nanjing Hydraulic Research Institute, Nanjing, China

2 Hohai University, Nanjing, China

*Address all correspondence to: qmzhong@nhri.cn

References

[1] Ministry of Water Resources of the People's Republic of China, National Bureau of Statistics of the People's Republic of China. Bulletin of First National Census for Water. Beijing: China Waterpower Press; 2013 (in Chinese)

[2] Ministry of Water Resources of the People's Republic of China. 2017 Statistic Bulletin on China Water Activities. Beijing: China Waterpower Press; 2018 (in Chinese)

[3] Zhang JY, Yang ZH, Jiang JP. An analysis on laws of reservoir dam defects and breaches in China. Scientia Sinica Technologica. 2017;**47**:1313-1320 (in Chinese)

[4] Dam Safety Management Center of The Ministry of Water Resources. Register of Dam Failures in China. Nanjing: Dam Safety Management Center of The Ministry of Water Resources; 2019 (in Chinese)

[5] Ru NH, Niu YG. Embankment Dam•Incidents and Safety of Large Dams. China Waterpower Press; 2001 (in Chinese)

[6] ASCE/EWRI Task Committee on Dam/Levee Breaching. Earthen embankment breaching. Journal of Hydraulic Engineering. 2011;**137**(12): 1549-1564

[7] Kirkpatrick GW. Evaluation guidelines for spillway adequacy. In: The Evaluation of Dam Safety, Engineering Foundation Conf., New York. 1977

[8] Soil Conservation Service (SCS). Simplified Dam-breach Routing Procedure. Washington, DC: US Department of Agriculture; 1981

[9] Hagen VK. Re-evaluation of design floods and dam safety. In: Proceedings of the 14th Congress of International Communicaton on Large Dams, Paris. 1982

[10] Singh KP, Snorrason A. Sensitivity of outflow peaks and flood stages to the selection of dam breach parameters and simulation models. Journal of Hydrology. 1984;**68**:295-310

[11] MacDonald TC, Langridge-Monopolis J. Breaching characteristics of dam failure. Journal of Hydraulic Engineering. 1984;**110**(5):567-586

[12] Costa JE. Floods from dam failures. In: Open-File Rep. No. 85–560. Denver: USGS; 1985

[13] Evans SG. The maximum discharge of outburst floods caused by the breaching of man-made and natural dams. Canadian Geotechnical Journal. 1986;**23**(3):385-387

[14] U.S. Bureau of Reclamation (USBR). Downstream hazard classification guidelines. In: ACER Tech. Memorandum No. 11. Denver: U.S. Department of the Interior; 1988

[15] Froehlich DC. Peak outflow from breached embankment dam. Journal of Water Resources Planning and Management. 1995;**121**(1):90-97

[16] Walder JS, O'Connor JE. Methods for predicting peak discharge of floods caused by failure of natural and constructed earthen dams. Water Resources Research. 1997;**33**(10): 2337-2348

[17] Xu Y, Zhang LM. Breaching parameters for earth and rockfill dams. Journal of Geotechnical and Geoenvironmental Engineering. 2009; **135**(12):1957-1969

[18] Pierce MW, Thornton CI, Abt SR. Predicting peak outflow from breached

embankment dams. Journal of Hydrologic Engineering. 2010;**15**(5): 338-349

[19] Thornton CI, Pierce MW, Abt SR. Enhanced predictions for peak outflow from breached embankment dams. Journal of Hydrologic Engineering. 2011;**16**(1):81-88

[20] De Lorenzo G, Macchione F. Formulas for the peak discharge from breached earthfill dams. Journal of Hydraulic Engineering. 2014;**140**(1): 56-67

[21] Hooshyaripor F, Tahershamsi A, Golian S. Application of copula method and neural networks for predicting peak outflow from breached embankments. Journal of Hydro-Environment Research. 2014;**8**(3):292-303

[22] Azimi R, Vatankhah AR, Kouchakzadeh S. Predicting peak discharge from breached embankment dams. In: E-Proc. 36th IAHR World Congress, Hague. 2015

[23] Froehlich DC. Predicting peak discharge from gradually breached embankment dam. Journal of Hydrologic Engineering. 2016;**21**(11): 04016041

[24] Mei SA, Chen SS, Zhong QM, et al. Parametric model for breaching analysis of earth-rock dam. Advanced Engineering Sciences. 2018;**50**(2):60-66 (in Chinese)

[25] Von Thun JL, Gillette DR. Guidance on Breach Parameters. Denver: Internal Memorandum, Bureau of Reclamation, U.S. Dept. of the Interior; 1990

[26] Froehlich DC. Embankment dam breach parameters revisited. In: Proceedings of the 1995 Conference On Water Resources Engineering, New York. 1995

[27] Cristofano EA. Method of Computing Erosion Rate for Failure of Earthfill Dams. Denver: US Bureau of Reclamation; 1965

[28] Zhong QM, Wu WM, Chen SS, et al. Comparison of simplified physically based dam breach models. Natural Hazards. 2016;**84**(2):1385-1418

[29] Fread DL. BREACH: An Erosion Model for Earthen Dam Failure. Silver Spring: National Weather Service, National Oceanic and Atmospheric Administration; 1988

[30] Brown RJ, Rogers DC. BRDAM Users' Manual. Denver: U.S. Department of the Interior; 1981

[31] Fread DL. DAMBREAK: The NWS dam break flood forecasting model. In: National Weather Service (NWS) Rep. Silver Spring: National Oceanic and Atmospheric Administration (NOAA); 1984

[32] Singh VP, Scarlatos CA. Breach erosion of earthfill dams and flood routing: BEED model. In: Res. Rep. Research Triangle Park: Army Research Office, Battelle; 1985

[33] Mohamed AAA, Samuels PG, Morris MW, et al. Improving the accuracy of prediction of breach formation through embankment dams and flood embankments. In: Proceedings of the International Conference on Fluvial Hydraulics (River Flow 2002), Louvain-la-Neuve, Belgium. 2002

[34] Morris MW, Kortenhaus A, Visser PJ. Modelling breach initiation and growth. In: FLOODsite Report T06–08-02, FLOODsite Consortium. 2009

[35] Wang P, Kahawita R, Mokhtari A, et al. Modeling breach formation in embankments due to overtopping. In: ICOLD Conference, Barcelona, Spain. 2006

[36] Temple DM, Hanson GJ, Neilsen ML, et al. Simplified breach

analysis model for homogeneous embankment: Part I, Background and model components. In: Proc. 25th Annual USSD Conference, U.S. Society on Dams, Denver. 2005

[37] Temple DM, Hanson GJ, Neilsen ML. WINDAM—Analysis of overtopped earth embankment dams. In: Proceedings of the ASABE Annual International Meeting, American Society of Agricultural and Biological Engineers, St. Joseph. 2006

[38] Hanson GJ, Tejral RD, Hunt SL, et al. Internal erosion and impact of erosion resistance. In: Proc., 30th U.S. Society on Dams Annual Meeting and Conference (CD-ROM), USSD, Sacramento. 2010

[39] Wu WM. Simplified physically based model of earthen embankment breaching. Journal of Hydraulic Engineering. 2013;**139**(8):837-851

[40] Chang DS, Zhang LM. Simulation of the erosion process of landslide dams due to overtopping considering variations in soil erodibility along depth. Natural Hazards and Earth System Sciences. 2010;**10**(4):933-946

[41] Chen ZY, Ma LQ, Yu S, et al. Back analysis of the draining process of the Tangjiashan barrier lake. Journal of Hydraulic Engineering. 2015;**141**(4): 05014011

[42] Wang L, Chen ZY, Wang NX, et al. Modeling lateral enlargement in dam breaches using slope stability analysis based on circular slip mode. Engineering Geology. 2016;**209**:70-81

[43] Chen ZY, Ping ZY, Wang NX, et al. An approach to quick and easy evaluation of the dam breach flood. Science China: Technological Sciences. 2019;**62**. DOI: 10.1007/s11431-018-9367-4

[44] Chen SS, Zhong QM, Cao W. Breach mechanism and numerical simulation for seepage failure of earth-rock dams. Science China: Technological Sciences. 2012;**55**(6):1757-1764

[45] Zhong QM, Chen SS, Deng Z. A simplified physically-based model for core dam overtopping breach. Engineering Failure Analysis. 2018;**90**: 141-155

[46] Zhong QM, Chen SS, Mei SA, et al. Numerical simulation of landslide dam breaching due to overtopping. Landslides. 2018;**16**(6):1183-1192

[47] Zhong QM, Chen SS, Fu ZZ. Failure of concrete face sand-gravel dam due to water flow overtops. ASCE's Journal of Performance of Constructed Facilities. 2019;**33**(2):04019007

[48] Zhong QM, Chen SS, Deng Z, et al. Prediction of overtopping-induced breach process of cohesive dams. ASCE's Journal of Geotechnical and Geoenvironmental Engineering. 2019; **145**(5):04019012

[49] Wu WM. Computational River Dynamics. London: Taylor & Francis; 2007

[50] Singh VP. Dam Breach Modeling Technology. Dordrecht, Netherland: Kluwer Academic; 1996

[51] Fread DL. DAMBREAK: The NWS Dam Break Flood Forecasting Model. Silver Spring: National Oceanic and Atmospheric Administration, National Weather Service; 1984

[52] Kesserwani G, Shamkhalchian A, Zadeh MJ. Fully coupled discontinuous Galerkin modeling of dam-break flows over movable bed with sediment transport. Journal of Hydraulic Engineering. 2014;**140**(4):06014006

[53] Wang Z, Bowles DS. Three-dimensional non-cohesive earthen dam breach model. Part 1: Theory and methodology. Advances in Water Resources. 2006;**29**(10):1528-1545

[54] Faeh R. Numerical modeling of breach erosion of river embankments. Journal of Hydraulic Engineering. 2007; **133**(9):1000-1009

[55] Wu WM, Wang SS. One-dimensional modeling of dam-break flow over movable beds. Journal of Hydraulic Engineering. 2007;**133**(1): 48-58

[56] Wu WM, Marsooli R, He ZG. Depth-averaged two-dimensional model of unsteady flow and sediment transport due to noncohesive embankment break/ breaching. Journal of Hydraulic Engineering. 2012;**138**(6):503-516

[57] Swartenbroekx C, Soares-Frazao S, Staquet R, et al. Two-dimensional operator for bank failures induced by water-level rise in dam-break flows. Journal of Hydraulic Research. 2010; **48**(3):302-314

[58] Li SC, Duffy CJ. Fully coupled approach to modeling shallow water flow, sediment transport, and bed evolution in rivers. Water Resources Research. 2011;**47**(3):1-20

[59] Cao ZX, Yue ZY, Pender G. Landslide dam failure and flood hydraulics. Part II: Coupled mathematical modelling. Natural Hazards. 2011;**59**(2):1021-1045

[60] Rosatti G, Begnudelli L. A closure-independent generalized roe solver for free-surface, two-phase flows over mobile bed. Journal of Computational Physics. 2013;**255**:362-383

[61] Rosatti G, Begnudelli L. Two-dimensional simulation of debris flows over mobile bed: Enhancing the TRENT2D model by using a well-balanced generalized roe-type solver. Computers and Fluids. 2013;**71**:179-195

[62] Juez C, Murillo J, Garcia-Navarro P. Numerical assessment of bed-load discharge formulations for transient flow in 1D and 2D situations. Journal of Hydroinformatics. 2013;**15**(4):1234-1257

[63] Juez C, Murillo J, Garcia-Navarro P. A 2D weakly-coupled and efficient numerical model for transient shallow flow and movable bed. Advances in Water Resources. 2014;**71**:93-109

[64] Swartenbroekx C, Zech Y, Soares-Frazao S. Two-dimensional two-layer shallow water model for dam break flows with significant bed load transport. International Journal for Numerical Methods in Fluids. 2013; **73**(5):477-508

[65] Guan M, Wright NG, Sleigh PA. 2D process-based morphodynamic model for flooding by noncohesive dyke breach. Journal of Hydraulic Engineering. 2014;**140**(7):04014022

[66] Kesserwani G, Shamkhalchian A, Zadeh MJ. Fully coupled discontinuous Galerkin modeling of dam-break flows over movable bed with sediment transport. Journal of Hydraulic Engineering. 2014;**140**(4):06014006

[67] Razavitoosi SL, Ayyoubzadeh SA, Valizadeh A. Two-phase SPH modelling of waves caused by dam break over a movable bed. International Journal of Sediment Research. 2014;**29**(3):344-356

[68] Marsooli R, Wu WM. Three-dimensional numerical modeling of dam-break flows with sediment transport over movable beds. Journal of Hydraulic Engineering. 2015;**141**(1): 04014066

[69] Abderrezzak KEK, Moran AD, Tassi P, et al. Modelling river bank erosion using a 2D depth-averaged numerical model of flow and non-cohesive, non-uniform sediment transport. Advances in Water Resources. 2016;**93**:75-88

[70] Cantero-Chinchilla FN, Castro-Orgaz O, Dey S, et al. Nonhydrostatic

dam break flows. II: One-dimensional depth-averaged modeling for movable bed flows. Journal of Hydraulic Engineering. 2016;**142**(12):04016069

[71] Cristo CD, Greco M, Iervolino M, et al. Two-dimensional two-phase depth-integrated model for transients over mobile bed. Journal of Hydraulic Engineering. 2016;**142**(2):04015043

[72] Cristo CD, Evangelista S, Greco M, et al. Dam-break waves over an erodible embankment: Experiments and simulations. Journal of Hydraulic Research. 2018;**56**(2):196-210

[73] Yan ZK. Research on Mechanism and Numerical Model of Tailing Dam Break due to Overtopping. Nanjing: Nanjing Hydraulic Research Institute; 2019 (in Chinese)

[74] Balsara DS. A two-dimensional HLLC Riemann solver for conservation laws: Application to Euler and magnetohydrodynamic flows. Journal of Computational Physics. 2012;**231**(22): 7476-7503